Why Don't Spinning Tops Fall?

Conversations With Curious Caroline

CHARLES DE LISI
Boston University

About the Author

Charles DeLisi is an American Biomedical Scientist and Dean *emeritus* of the College of Engineering at Boston University. From 1999 until his retirement in 2024 he was the Metcalf Professor of Science and Engineering. He is currently Distinguished Visiting Professor at Boston University's Center for Data Sciences and Research Professor in the College of Engineering. In addition to authoring or co-editing ten books, he, his students and collaborators have published more than 300 research articles in biophysics, cancer genetics, immunology and computational science. In 2001 he was awarded the Presidential Citizen's Medal for seminal contributions to the initiation of the Human Genome Project. In conferring the honor President Clinton said: "Just as Lewis and Clark set forth to explore a continent shrouded in mysterious possibility, Charles DeLisi pioneered the exploration of a modern-day frontier, the human genome... [his] imagination and determination helped to ignite the revolution in sequencing that would ultimately unravel the code of human life itself."

Figure Credits

Cover
Spinning tops made by David Earle.
https://commons.wikimedia.org/wiki/File:Snap_Top_and_Trompo.jpg

This file is licensed under the Creative Commons Attribution-Share Alike 4.0 International license.

Conversation 2

Figure 1. **DALL-E** conception of a pre-teen in her room enjoying her violin

Figure 2. A typical tuning fork. https://commons.wikimedia.org/wiki/File:Diapason.jpg

Figure 4. Cartoon of compressions (columns of dense dots representing air molecules) and rarefactions (sparse dots) that transmit energy https://images.app.goo.gl/7vHoCAbKgapf6jQh6

Figure 6. Sine wave with stars

Figure 7. The anatomy of the ear, Lars Chittka; Axel Brockmann Creative Commons Attribution 2.5

Figure 10. **Dall-E** rendition of Caroline musing over why the sounds of music are so moving

Conversation 3

Figure 1. Left: A general overview of a precessing top. http://hyperphysics.phy-astr.gsu.edu/hbase/top.html.

Conversation 4

Figure 1. A monochromatic rainbow after a rain shower just as the Sun is rising. Licensed under the CC-BY-SA-4.0 https://commons.wikimedia.org/wiki/User:Smurrayinchester

Figure 2. https://commons.wikimedia.org/wiki/File:Refraction_of_light_at_plane_surface_-_rarer_medium_to_denser_medium.png

Figure 3. Light bending as it enters a medium of higher index of refraction. https://upload.wikimedia.org/wikipedia/commons/0/01/Huygens-Refraction.jpg Creative Commons Attribution-Share Alike 4.0 International license.

Figure 4. This image has been created during "DensityDesign Integrated Course Final Synthesis Studio" at Politecnico di Milano, organized by DensityDesign Research Lab in 2015. Credits goes to Ludovica Lorenzelli, CC BY-SA 4.0, via Wikimedia Commons

Conversation 5
Figure 1. Chat GPT 4.o generated figure

Figure 2. Induced current by changing magnetic field. Juancarcole, CC BY-SA 4.0, via Wikimedia Commons https://commons.wikimedia.org/wiki/File:Experiencia_de_Faraday.gi

Conversation 6

Figure 1. X-rays appear to emanate from Superman's eyes. https://picryl.com/media/x-ray-vision-3af1d8

Figure 2. One of Roentgen's photographic plates (https://www.aps.org/apsnews/2001/11/roentgen-discovery-X-rays

Figure 3. Plot of the number of nations with nuclear weapons vs year.

Figure 4. Photon knocking an electron from its orbit. https://study.com/academy/lesson/lewis-structures-single-double-triple-bonds.html

Figure 5. A four nucleotide segment of a single strand of DNA. https://commons.wikimedia.org/wiki/File:DNA-Nucleobases.svg

Figure 6. **Dall-E** rendition of New York's Central Park in winter.

Figure 11. A cartoon of the DNA double helix (https://brainly.com/question/8998157), left and the diffraction pattern from which information about its structure can be obtained https://en.wikipedia.org/wiki/Photo_51#/media/File:Photo_51_x-ray_diffraction_image.jpg PA1707, CC BY-SA 4.0, via Wikimedia Commons

Conversation 7

Figure 1. The global land and sea average temperature relative to the average temperature between 1951 and 1980. https://climate.nasa.gov/vital-signs/global-temperature/?intent=121 NASA/GISS

Figure 2. Global temperature changes during the last 60M years. http://www.alpineanalytics.com/Climate/DeepTime.html

Figure 3. A simplified summary of carbon exchange between four of its reservoirs: land, atmosphere, ocean and deep ocean. http://earthobservatory.nasa.gov/Features/CarbonCycle/adapted.fromU.S.DOE,Biological.and.Environmental.Research.Information.System

Figure 4. Keeling curve https://keelingcurve.ucsd.edu

Figure 5. 100 K year temperature cycles. Source: J. R. Petit and others, "Climate and Atmospheric History of the Past 420,000 Years from the Vostok Ice Core. Antarctica." Nature 399 (Tune 1999): 429-36.

Conversation 8

Figure 1. Rectangular conducting wire rotating at angular velocity ω in a radially outward pointing magnetic field CC BY-SA 4.0 https://commons.wikimedia.org/wiki/File:Spindle.PNG

Figure 3. Step-up transformer Transformer3d col3 /BillC at English Wikipedia / CC BY-SA 3.0) https://en.wikipedia.org/wiki/File:Transformer3d_col3.svg

Conversation 9

Figure 1. The main elements of a neuron. http://id.nlm.nih.gov/mesh/D001369
http://purl.org/sig/ont/fma/fma67308

Figure 2. When an action potential reaches an axon terminal it stimulates the release of neurotransmitters which bind to receptors on the dendrites of adjacent cells. Image by Thomas Splettstoesser / CC BY-SA 4.0

Figure 3. An outline of the pathways from the retina to the primary visual cortex. Visual Pathway from https://en.wikipedia.org/wiki/Visual_system#/media/File:Human_visual_pathway.svg

<u>Drawings by the author</u>

Conversation 1 Figures 1 and 2
Conversation 2, Figure 4, Figure 7, Figure 8
Conversation 3 Figure 1, right panel
Conversation 6 Figures 7 -10
Conversation 8 Figure 2, Figure 4, Figure 5
Conversation 9 Figure 4, Figure 5, Figure 6, Figure 7, Figure 8

Dedication

To the genius of free higher education and to The City College of New York in particular.
To Sidney Moskowitz, BSEE, CCNY, 1939

Acknowledgements

I am deeply indebted to Noreen Vasady-Kovacs for insightfully line editing the entire manuscript as it evolved through various versions and to Zhiping Weng for reading and commenting on the final version of the manuscript, and critically appraising the content of an earlier version. I am also grateful to Dustin Holloway and Simon Kasif for helpful suggestions on Conversation 9.

Preface

"Mommy, why is grass green?"

Mother: "That's a good question; ask your father."

"Daddy, what holds up the moon? And my spinning top — just because it spins it doesn't fall?"

Father: "Yes, it does seem strange now that you ask. Tops seem to defy gravity— maybe your mother has an explanation."

"Mommy, Daddy, do you mind that I'm asking all these questions"

Parents: "…of course not Caroline; how else are you going to learn?"

For many adults, curiosity about the world fades over time. The mysteries surrounding us become so embedded in daily life that they no longer register as mysteries — until someone asks a question. Such questions, often posed by children, are jarring and almost embarrassingly difficult to answer. Not surprisingly, they frequently go unanswered and eventually they cease altogether.

This book seeks to rekindle that sense of curiosity, addressing questions that might otherwise remain unexplored. While the idea of tackling everyday scientific and technological mysteries is not new, my approach here is different. Rather than presenting a list of questions with concise answers, I've chosen to frame these explorations within a narrative structure.

Initially, I intended to write a straightforward question-and-answer book, offering succinct explanations for common phenomena. However, as I began writing, I found myself drawn to the idea of conversations. These dialogues involve not just questions and answers but also the interplay of perspectives among family members and friends. They mimic the organic flow of real-life discussions, where one question sparks another, and understanding deepens through shared exploration. Consequently, the book unfolds as a series of narratives that reflect the natural curiosity of its central character, Caroline, as she matures from a 13 year-old high school freshman to a 17 year-old senior. Her enthusiasm drives free associations, connecting seemingly disparate ideas with remarkable agility. The discussions often diverge into unexpected territory, from ethical debates to poetry, literature, and music, while always returning to the central theme of science. This approach mirrors the interconnectedness of knowledge and the unpredictable nature of unstructured conversation.

It's important to emphasize that this book is not a textbook. Unlike a traditional text, its chapters do not form an interdependent sequence, nor do they confine themselves to a single discipline. Each chapter is a self-contained exploration, guided by Caroline's curiosity. This unstructured approach allows for spontaneity but also requires readers to engage actively, connecting the dots as they progress.

Writing Caroline posed two major challenges. The first was recalling the questions that once seemed so pressing. With time and familiarity, even the most intriguing mysteries can lose their luster. The second was grappling with the limits of human understanding. How much of what we "think" we know is truly understood? While this is a profound question, beyond the scope of this book, I touch on it briefly to establish a standard for what constitutes a satisfactory explanation, and to address the question of when to stop asking why in response to an answer.

The answers presented here often reflect well-established knowledge, particularly in engineering and technology. However, simplifying complex concepts for an audience ranging from late grade school to early college inevitably leads to some level of superficiality. To address this, I've included references after each section to encourage deeper exploration. My hope is that readers will pursue these topics further, embracing the layered complexities that make even simple phenomena fascinating.

Although I've aimed to make the science accessible to a thoughtful high school student, the narrative itself often operates at a level more suited to an intelligent adult. This choice reflects my belief that people learn best when they actively engage with material, seeking out unfamiliar words or concepts to enhance their understanding. Consequently, some sections may require a dictionary, Google, or a chatbot for full comprehension.

Finally, I've chosen to focus on a limited number of topics, presented as 69 declarative statements across nine chapters. While I could have included twice as many, this seemed a reasonable scope for an experimental book. Caroline, after all, is well on her way to a distinguished career and is fully capable of conducting her own research.

It is my hope that this book will inspire teenagers—and especially young women, who remain seriously underrepresented in the hard sciences—to explore the mysteries of the commonplace through science, technology, engineering, and mathematics (STEM). More broadly, I hope that *Caroline* will encourage readers to rediscover the joy of seeing the world through the fresh eyes of an adolescent with a childlike curiosity who can't stop asking, "Why?"

Table of Contents

Conversation 1 1

Blood, Straws and Pressure 1

Newton's Law of Gravity Applies Everywhere in the Known Universe 1

Torricelli Invents the Barometer and Determines that Air Has Weight 7

The Weight of the Atmosphere Pressing Down on the Angstroms' Kitchen Table is Nearly 59,000 Pounds 11

Fluids Can Be Made to Defy Gravity by Manipulating Pressure 13

Atmospheric Pressure is 6 - 8 Fold Greater Than the Arterial Pressure Exerted by Blood in a Healthy Adult 19

Conversation 2 22

Timbre and the Sounds of Music 22

Caroline's Home is Steeped in Beautiful Noise 23

Waves Permeate the Universe 27

All Waves Transport Energy and Have Common and Unique Properties 27

A Musical Note Has a Complex Structure 28

Tuning Forks Generate Only the Fundamental Frequency of a Note 29

Sound Moves Forward but the Molecules of Air Through Which it Moves Only Vibrate 32

Frequency, Wavelength, and Speed Are Connected 34

Sound Propagation in Water is More Complex Than it is in Air 35

The Human Auditory System Can Detect Pressures a Million Times Smaller Than an Ant Exerts on a Hard Surface.. 37

Visible Light and Other Electromagnetic Waves Differ Only in Their Frequency... 43

A 2-foot Long Piano String Produces a Middle C, Which Has a 4-Foot Wavelength 47

Some Waves Stand Still.. 50

The Same Note Played on Different Instruments Sounds Different Because Different Instruments Produce Different Harmonic Spectra.. 56

Conversation 3.. 59

Planes, Tops, Gyros and the Apparent Defiance of Gravity 59

Newton's Third Law and Pressure Differentials Help Keep Kiloton Planes in the Air 59

Daniel Bernoulli: A Legacy of Scientific Genius.. 64

Birds and Planes: Different Objectives, Same Physics...................................... 65

A Spinning Top Generates a Gravitational Torque that Changes the Direction of its Angular Momentum, Causing it to Precess.. 67

Airport Security Imaging Is Not a Serious Health Hazard................................. 76

Conversation 4.. 80

Color, Rainbows and Mirages 80

Understanding Why a Blue Sky Turns Reddish Orange in the Evening Involves Both Physiology and Physics... 82

Caroline's Mother is Darker than her Father Because her Skin has a Higher Concentration of Light-absorbing Molecules 84

The Search for Understanding Can Lead to an Infinite Regress of Causes and Whys 84

A Rainbow is Caused by the Refraction, Reflection, and Dispersion of Light Within Water Droplets 89

A Mirage is Caused by the Refraction of Light in Layers of Air with Varying Temperatures and Densities 91

Conversation 5 97

Light: We See so Little 97

We're Flooded by Light We Don't See 98

Everyday Devices Such as TV Remote Controls and Garage Door Openers are Controlled by Light With Vastly Different Frequencies 101

The Time That Elapses Between Hearing and Seeing Visible Light Reveals the Distance of an Approaching Storm 103

The Air Surrounding Lightning s Hotter Than the Surface of the Sun 104

Lightning Creates Shock Waves 105

The Frequency of an Electromagnetic Wave Has a Profound Effect on Its Behaviour 107

An Electric Current Produces a Magnetic Field 110

Maxwell and Faraday Developed the Principles of Electricity and Magnetism that Transformed Society 112

Microwaves Cause Water and Other Polar Molecules in Food to Vibrate and Rotate, Producing Heat Through Friction and Molecular Motion 116

Conversation 6 118

Seeing Through Objects 118

Superman Would Have Been More Realistic with Infrared Vision 119

Roentgen Discovered X-rays and Demonstrated Their Potential for Enabling a Medical Revolution 123

Einstein's Paradigm-Shifting Theory of the Photoelectric Effect Laid the Foundation for New Industries 125

X-rays Can Display Anatomy Because They Are Absorbed More Effectively by Bone Calcium Than by Soft Tissue 131

Thomas Young's Double Slit Experiment Established the Wave Nature of Light .. 141

Light Behaves as Both Wave and Particle 143

X-rays Can Help Visualize Proteins and DNA 145

The Electronic Structures of Glass and Wood Explains Why Light Can Penetrate One But Not the Other 146

Conversation 7 148

Our Planet's Changing Climate 148

The Earth is an Organism—and it Has a Fever 148

A Small Increase in Average Global Temperature Can Radically Disrupt the Planetary Ecology 149

Several Successive Years of Declining Temperature Does Not Mean the Fever is Subsiding 152

Natural Processes as Well as Human Activities Drive Long-Term Temperature Trends 154

Natural Changes in Climate Are Easily Distinguished From Anthropogenically Driven Change 154

Greenhouse Gases Absorb Heat 157

Infrared Radiation Emitted by the Earth is Absorbed and Reemitted Many Times by Greenhouse Gases Before Escaping the Planet 160

In the Absence of Greenhouse Gases, the Average Surface Temperature of the Earth Would be ~ -18°C 161

The Atmospheric Concentration of CO₂ Cycled Between 125 ppm and 290 ppm for 800 Thousand Years Then Spiked to 420 ppm During the Last 150 Years 167

Atmospheric Conditions that Existed Millions of Years Ago Can be Inferred by Analyzing Air Bubbles Trapped in Ice 172

Multiple Lines of Evidence Indicate That Modern Climate Change is Anthropogenically Driven 173

Feedback Has a Pronounced Effect on the Rate at Which the Planetary Temperature is Increasing 175

If CO₂ Emissions Ceased Immediately, the Global Temperature Would Increase by ~ 0.6°C Before Stabilizing 177

Many Innovative Clean Energy and Atmospheric Cleansing Technologies Have Been Proposed but None Have Been Implemented at Scale 180

Conversation 8 182

Clean Energy and the Generation of Electricity 182

A Small Increase in the Concentrations of Heat Absorbing Gases can Increase the Average Global Temperature 184

Wind Turbines Convert the Mechanical (Kinetic) Energy of Wind into Electricity 187

The Kinetic Energy of Moving Water and the Heat Released by Nuclear Fission Also Produce Electricity by Driving Turbines 198

Photocells use Photons to Excite Electrons in Doped Semiconductors, Creating an Electric Current Through the Photovoltaic Effect 199

Conversation 9 203

Artificial Intelligence (AI) 203

Machine Learning is One Type of AI and Deep Neural Networks Are One Type of Machine Learning 204

The Impact of AI: Both Profoundly Positive and Profoundly Negative 206

The 86 Billion Neurons of the Central Nervous System Are Clustered into Large Communication Networks 212

Artificial Neural Networks Learn by Modulating the Interactions Between Neurons, Just as Biological Networks Do 219

Large Language Models (LLMs) Are Neural Nets That Can Respond to Long, Complex Queries With Proper Syntax and Semantics in Many Domains of Human Knowledge 228

Epilogue 232

Caroline's Notebook 236

References 243

Index 255

Conversation 1

Blood, Straws and Pressure

Caroline Angstrom asks simple questions about blood pressure and the stock market, but her parents, Mary and Don, answer with an extended discussion about pressure while placing on hold a conversation about the unrelated field of financial markets. She is stunned when she's told that air has weight, and surprised even more when she learns about gravity defying fluids. Finally, Caroline receives the answer to the question she asked weeks earlier, when her Aunt Zoe explains blood pressure and those strange numbers.

Newton's Law of Gravity Applies Everywhere in the Known Universe

It's six p.m. and Mary Angstrom has just arrived home from Packard Analytics, a Wall Street firm that provides investment advice. She looks more frazzled than usual, and her husband Don is concerned, as is her daughter, Caroline.

"I've had a difficult day," she says, responding to Don's disquiet as she finds her way to the sofa. "The stock market dropped another half a percent — it was the tenth drop in prices in the last thirty days and it comes in a year when the market has been both volatile and bearish."

Caroline: "What does bearish mean?"

"I thought Packard Analytics was doing reasonably well," Don says with surprise, being too focused on his wife for Caroline's question to penetrate.

Mary: "It's doing better than most funds, but that's not saying much. Sometimes I feel like I'm ready to retire."

Don: "Well, retiring at the age of thirty-five is great if you can do it."

Mary: "Yes, nice work if you can get it, as the song goes. Even my blood pressure's up." Mary turns to her husband and after a pause says with a hint of good-humored and loving sarcasm—"I just took it before I left my office with that romantic sphygmomanometer you gave me for my birthday."

"What was your blood pressure?" Don asks with concern.

1

"One-forty over ninety—the systolic and diastolic pressures are both high."

Caroline, whose curiosity caused her to lose sight of Mary's condition, impatiently interjects, telling her mother that she doesn't understand a word they're saying. "I mean, I know Packard Analytics is the company you work for, but what's a stock market—and what's a sfigo…"

Don immediately intervenes: "Sphyg mo man ometer—a device that measures blood pressure."

Caroline (now with perfect diction): "Sphygmomanometer—but what's blood pressure? And what do those numbers mean, one-forty over ninety? And you still haven't told me what a stock market is."

Don, who heard all of Caroline's questions: "OK, let's take one thing at a time. I'm sure you know what a market is."

Caroline: "Yes, it's a place where you buy stuff."

Don: "Yes, a market is a place where people buy and sell commodities, trade goods and services—and a stock market gives people a way to trade, to buy and sell, parts of companies. Someday when the weather's unpleasant, we can set aside time to discuss buying, selling and trading, but for now let's focus on blood pressure."

Mary, who is now starting to feel a bit better, turns to her husband and asks if he wants to take a first shot at trying to explain blood pressure.

Don: "Why don't we do it together on Saturday when we'll have more time—you know our daughter, she can be exhausting, and besides we still haven't had dinner, and I'm also a bit tired."

It's Saturday morning just after breakfast, and Mary, Don and Caroline move to Mary's study where she begins to explain those mysterious blood pressure numbers.

Mary: "In order to understand what those numbers mean, you need to know the meaning of pressure, and to understand pressure you need to understand force, and to understand force you need to know the difference between mass and weight.

"*Mass is the amount of material in an object, for example, the number and types of molecules. Weight is the force of attraction between two masses*, e.g. your weight is the gravitational force of attraction between the mass of your body and the mass of the Earth. Have you heard of Newton's law of gravitation?"

Caroline: "I've heard it mentioned but I don't know anything about it."

Mary: "I'm sure you'll learn about it next year when you enter high school. *Newton's law of universal gravitation says that the force between any two objects is proportional to the product of their masses and inversely proportional to the square of the distance between them.*

The remarkable thing about it is its universality: it applies to any two masses in the Universe—not just between you and the Earth, but also between the Earth and the Moon, the Moon and the Sun—literally between anything, even between you and your glass of milk, but that force is so small compared to the gravitational attraction of the Earth that you don't notice it."

"Yes, Mom, it doesn't take much for me to resist the pull of a glass of milk," Caroline quips with a winning smile.

Mary continues, ignoring her daughter's suggestion that she doesn't care for milk. "There are many other kinds of force—I'm sure you're familiar with magnetism, which you can feel when you bring two magnets close together."

Caroline: "What about when I push against this table? Am I exerting a force on it?"

Mary: "Yes, that's an example of a mechanical force."

Caroline: "Even if the table doesn't move?"

Mary: "Yes, even if it doesn't move. That's not so strange. When you remain perfectly still, the Earth is still exerting a force on you, right?"

Caroline: "I guess."

Don interjects: "You're touching on the difference between force and work, and I'd like to avoid digressing too much. Can we save that discussion for another day?"

Caroline: "OK."

Don continues: "Pressure is closely related to force. Scientists talk about pressure—force divided by the area on which it acts. For example, the pressure that your history book exerts on the table is the weight of the book divided by the area in contact with the table. If a book that weighs five pounds and has an area of ninety-six square inches (sq. in.) is placed flat on the table, how much pressure does it exert?"

Caroline: "Five divided by ninety-six?"

Don: "Which is?"

Caroline: "Can I use my calculator?"

Don: "No. Suppose the contact area was one hundred square inches (sq. in. or in²), what would the pressure be?"

Caroline: "Five divided by one hundred, or one-twentieth of a pound per square inch. So it's a little more than zero point zero five pounds per square inch? Now I see how to make the estimate. If the book has an area of ninety-six square inches, the pressure would be closer to one-nineteenth of a pound per square inch (five divided by ninety-five)."

Don: "Yes, you've got the idea."

Mary: "By the way, and this is important, the word pound as defined by *the British System of scientific standards refers to mass, not force. The unit of force (weight) is the pound-force (lbf). However, lb (pound) and lbf tend to be used interchangeably*."

Caroline: "That really *is* confusing."

Caroline then retrieved a notebook that she had recently bought to keep a record of discussions that confused her. The notes proved to be very helpful and she would invariably return to them and augment them as her understanding slowly developed.

Mary continues: "It can be, but it usually isn't because context can clarify. The possibilities for confusion, however, go beyond ambiguity. The USA and the United Kingdom are among the few countries in the world that use the so-called Imperial system of units—most of the rest of the world uses the metric system, which I was planning to talk to you about later. For now, just keep in mind that if you weigh yourself in France or Italy, or any number of other countries, the scale will display kilograms, which is a unit of mass."

Caroline: "And in Britain and the U.S. does the scale show your weight or your mass?"

Don: "All scales measure the gravitational force exerted by your body, which is your weight. However, they are calibrated to display your mass. For instance, in Europe, scales measure your weight in newtons but are calibrated by dividing the gravitational field intensity which is 9.8 meters/second^2 (m/s^2) to display your mass in kilograms. On the other hand, in the U.S. and Britain, the scale divides by thirty-two feet per second squared."

Don continued by noting that as long as everyone agrees on the meaning, the label—pounds, kilograms, or anything else—doesn't really matter. Then he added with an impish grin, "A pound by any other name would weigh the same."

Mary (rolling her eyes): "Really, Don?"

Caroline, to her mother: "What's Dad talking about?"

Mary: "Your father is alluding to a famous line from one of Shakespeare's most popular romantic tragedies, Romeo and Juliet, in which Juliet laments her family's opposition to her involvement with Romeo because of a long-standing feud between her family and his, and not because of substance. It's been a long time since I read the play, but who can forget the famous couplet:

What's in any name? That which we call a rose.

By any other name would smell as sweet."

Don: "I thought I was being clever as well as humorous—there is a connection."

Mary: "Please ... I think you owe Shakespeare an apology."

Don: "I was of course only trying to add some levity—Juliet's words capture so precisely and presciently the tragedy that's about to unfold as the result of conflating descriptors and substance. Shakespeare, as we both know, was no ordinary genius, and the play has a profound message, whereas my discourse on units is not, to say the least, nearly as deep."

Caroline: "The play sounds fascinating and I can't wait to read it, but I have to say that sometimes the two of you are more childlike than I am. Anyway, I'd like to return to the discussion of weight and mass so I can be sure I understood correctly: whether I weigh myself in Europe or the United States I'm finding my mass, but the units are different, so the numbers are different."

Don: "Yes, you've got it—but enough, let's move on."

Caroline was so caught up in the distinctions that she didn't fully assimilate what her father just said about moving on, and asked what she would weigh on the Moon.

Mary: "You know that the gravitational force between two objects, such as between you and the Earth or you and the Moon, depends on the product of your masses. That means your weight on the Earth, or the force between you and the Earth, would be proportional to your mass times the Earth's mass, and your weight on the Moon would be proportional to your mass times the Moon's mass—so of course you would weigh a lot less on the Moon."

Caroline: "I know all that, but what would I weigh?"

Mary: "If you were using a scale properly calibrated to the Moon's gravity, your scale would display the same number that it does on Earth because your mass is independent of whether you're on the Earth or on the Moon or anywhere else in the Universe. However, your weight would be different, but that's not what scales display."

Caroline: "OK, I think I understand, but I still want to know what my weight would be on the Moon."

Don: "Don't you think that you should be able to figure that out for yourself?"

Caroline: "No, because I don't know what the distance would be between me and the Moon, I mean I'd be standing on the Moon's surface, but the distance can't be zero."

Don: "That's an important point. I'll tell you what to use but the explanation will probably have to wait until you study mechanics in college."

5

Caroline: "Try me; I'd like something to think about."

Don: "It has a simple answer that's difficult to understand—you can think of the mass of the Earth as being entirely at its center, so since the Earth is essentially spherical and you're standing on its surface, the distance you need is the radius of the Earth."

Caroline: "You were right, I don't understand why you can consider the entire mass of a spherical object to be at its center."

Don: "That's OK, even Newton took a while to prove it, so do give it some thought—but for now let's continue with our more pedestrian discussion of atmospheric pressure."

Caroline: "OK, I've taken a bunch of notes and I'll get back to you when I think I have an answer."

Torricelli Invents the Barometer and Determines that Air Has Weight

Mary: "The atmosphere, the air all around us, which we breathe, also exerts pressure. *At sea level when the air is dry, and the temperature is 32 Fahrenheit (32 °F), the atmospheric pressure in the Imperial system of units, that is the system used in Great Britain and the United States, is 14.7 pounds per square inch*—where for simplicity pounds per square inch is often abbreviated as psi or lbs./in²."

Caroline: "Dry air at sea level and 32F? Does atmospheric pressure depend on altitude, air moisture and temperature?"

Mary: "Yes, it does. For example, atmospheric pressure is about 12% lower at 1000 m, and moist air is less dense than dry air."

Caroline: "I don't understand."

Mary: "I can see we're going to get into a series of whys that will take us too far afield, but I'll say a couple of things just to provide some context. First, moist air is less dense than dry air because water molecules are lighter on average than the molecules they replace."

Caroline: "Water replaces molecules?"

Mary: "You don't stop! The answer is yes, because the number of molecules of air per unit volume at a given temperature is pretty much fixed. I'm sure you can figure out the answer to any follow-up questions you have with a little bit of research and thought.

With respect to a quantitative relation between pressure and altitude, let's add that to our rainy Saturday list. One question always leads to another, and we'll never get to answer your questions about blood pressure if we digress too much."

Caroline: "It's hard to imagine how you can weigh air—how do you know that the pressure is 14.7 lbs. per square inch?"

Don starts to explain by drawing a picture of a simple barometer (Figure 1), then realizes that he should back up a bit and say something more about the units used for specifying mass and distance.

"I know we've already discussed units but there are still a few gaps that need to be filled in and I'd also like to make sure everything is crystal clear before we delve into the physics of a barometer — so here we go again," Don sighs with a hint of exasperation and frustration.

Don goes into a lengthy explanation about units, which Caroline writes into her notebook, and he concludes by saying that they'll stick with the metric system, or SI units, which is less complicated than the Imperial system. In addition, since Caroline already knows that the meter is a unit of length and the kilogram (kg) is the unit of mass, the only additional piece of information needed is the unit of force, which is the newton (N). More specifically, a one kilogram mass subject to a force of one newton will accelerate at a rate of 1 meter/sec² (1 m/s²).

Caroline: "What does SI stand for?"

Don: "It's a French abbreviation for système international d'unités, and in that system distance is measured in meters, mass is measured in kilograms and time, as usual, is measured in seconds—so you'll often see MKS.

"Now we should be able to talk about *pressure, which has units of N/m²*, but people couldn't leave well enough alone and *gave N/m²* a name—the *Pascal (Pa)*, in honor of Blaise Pascal the French mathematician, physicist, and philosopher who lived at the same time as Galileo and Milton, and generalized the work of Torricelli, among many other contributions. Pascal was a remarkable intellect, a child prodigy who wrote his first treatise on geometry at the age of 16, made major contributions to a number of areas of science, engineering and philosophy, was one of the greatest writers of prose in the history of France—and died tragically at the age of 39. He obviously deserved to be honored, as he was in many ways, including by the unit of pressure bearing his name.

"Before moving on I'll leave you with one more factoid that illustrates how the same quantity can have strikingly different numerical values depending on the system of units. You already know that atmospheric pressure in the British system is 14.7 lb./in², but what you probably haven't heard is that it's about 10^5 N/m² or 10^5 Pa in SI units."

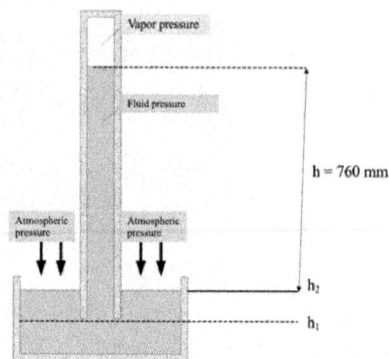

Figure 1. The inverted test tube is initially filled with mercury and partially submerged in a basin of mercury. The external pressure at h_1 is due to atmospheric pressure plus the pressure due to mercury between h_2 and h_1. Since the initial pressure inside the column at h_1 exceeds the pressure outside the column at h_1, mercury will leave the column until the internal and external pressures equalize. At that point the pressure of mercury above h_2 must equal the atmospheric pressure on the mercury in the basin. Since the column equilibrates when that distance is 760 millimeters (mm), atmospheric pressure is able to sustain a column of mercury 760 mm high (approximately 30 inches).

Caroline: "Yes, that's a striking difference and it could create difficulties in developing an intuitive feeling for the magnitude of physical quantities—which I guess emphasizes why it's best to stay with just one system of units. We've started to learn about SI units in science, and my teacher says that it seems more natural than the British system because it's based on multiples of ten—ten millimeters in a centimeter, one hundred centimeters in a meter, one thousand meters in a kilometer and so on. That's much simpler than the British system that we use here in the U.S.: twelve inches in a foot, three feet in a yard, five thousand two hundred eighty feet in a mile, not to mention ounces and pounds—or should I say pounds-force. It seems strange that we don't use the metric system."

Don: "I've heard some congressmen say they don't understand it."

Caroline: "But it seems so simple."

Don: "Yes, it is, but now let's get back to measuring atmospheric pressure.

"The story begins with Evangelista Torricelli who was a student of Galileo's. Like his teacher, Torricelli made a number of important contributions to physics and mathematics, the most famous being the hypothesis that air has weight. His own words say it best: 'We live submerged at the bottom of an ocean of the element air, which by unquestioned experiments is known to have weight.' The unquestioned experiments he refers to, which verified the startling announcement, were barometric measurements made with an instrument that he himself invented."

Pointing to the figure, Don explained that at equilibrium the upward pressure at the base of the column due to external forces is the pressure of the atmosphere plus the pressure due to the weight of mercury between h_2 and h_1—and the downward pressure at the base of the column is the pressure due to mercury between h_1 and h_2 plus the pressure due to mercury occupying length h of the column above h_2. Since the pressure between h_1 and h_2 is the same inside and outside the column, the pressure of mercury in the column which is above h_2 must be the same as atmospheric pressure. So atmospheric pressure is determined simply by determining the height, h, of mercury in the column relative to the surface of mercury in the basin.

Caroline: "OK, but what is the pressure of a column of mercury of height h?"

Don: "Figure it out. What's the definition of pressure?

Caroline: "Pressure is defined as weight per unit area."

Don: "So let's say the cross-sectional area of the column is a, and ρ_m is the density of mercury, what is an expression for the pressure? You can begin by calculating the total mass in the column."

Caroline: "The definition of density is mass per unit volume, so the total mass is $\rho_m ah$. And you told me that weight is mass multiplied by the gravitational acceleration, g, so the pressure is

$$\rho_m gha/a = \rho_m gh."$$

Don: "Very good. Notice that the only variable in this expression is h: ρ_m and g are constant, so h is sufficient to specify pressure."

Caroline: "OK, that's very interesting. Now I see why *meteorologists might say that the atmospheric pressure is 760 millimeters (mm). Even though the millimeter is not a unit of pressure, people understand that the height is multiplied by $\rho_m g$.* I can see why it's so difficult to stick to a single system of units—the height of a column of mercury is just another unit of pressure and a confusing one at that, since mm is a measure of length, but at least it's an SI unit."

Caroline continues: "Anyway, this is all very clever, I mean the way atmospheric pressure can be determined so simply, and described so easily—but isn't it surprising that no one ever thought of it before Torricelli?"

Don: "Caroline, the more you learn about human progress, the more you'll encounter new ideas that seem obvious after they're articulated by someone, but are not so obvious before that. You've probably heard the story of Columbus's egg."

Caroline: "No, I haven't."

Don: "I'll leave it to you to look up the Columbus egg story, but I'll let you in on wisdom that you can't look up—it was conveyed by people who never put their wise words to paper because some ideas are more readily communicated verbally. I had a physics teacher when I was in college who used to say that everything in the history of physics is either trivial or wrong, and your uncle Zack had a post-doctoral advisor who once told him everything is easy—after it's explained. Both statements are of course exaggerations, but exaggerations are sometimes powerful ways to convey a nearly universal truth."

Caroline: "I guess having a good teacher is a big advantage."

Don: "Yes, a teacher who can explain subtle concepts in an intuitive and clear manner will help most of us learn and understand much more readily than otherwise. But of course there is that small fraction of uncommon geniuses who not only need no help, but quickly go beyond the cumulative understanding of those who came before."

The Weight of the Atmosphere Pressing Down on the Angstroms' Kitchen Table is Nearly 59,000 Pounds

Don knew that Caroline had learned how to calculate the area of a circle, but he was curious about her ability to apply what she had learned.

"If the glass is a circular cylinder with a diameter of three inches, how much atmospheric weight would be on the surface of the milk?"

Caroline thinks a bit, then says: "The area is πr^2 and r = 1.5 inches and the weight is 14.7 pounds per square inch—so it's 3.14 × 1.5 × 1.5 × 14.7 pounds. Can I borrow your phone?"

Her father hands her his phone.

Caroline used a calculator that her father had downloaded to find the answer: 103.8 pounds. "That's incredible; it's more than I weigh. This is really neat—I can figure out stuff that I could never have even guessed if I didn't do the calculations."

"Well if you think that's neat then you'll love science: there are so many startling discoveries that have been made just by calculating."

Caroline then decided to calculate the atmospheric weight on the entire table and was shocked when she found that the weight of the atmosphere on the table is nearly 59,000 pounds: "That can't be right: why wouldn't the table collapse?"

Mary explained that the air pressure is the same on all parts of the table—the air on the bottom of the table pushes up with the same pressure that the air on the top of the table pushes down.

Caroline: "I'm confused. Isn't the weight on the top of the table the result of all the air from the tabletop to the top of the atmosphere?"

She then continued, without waiting for an answer. "The air on the bottom of the table only extends to the floor. How can it be pushing up with the same force?"

Mary: "When you say from the tabletop to the top of the atmosphere, it sounds like you're thinking of air as a continuous substance like a fluid. There's a better way to think about it.

"Remember that air is composed of molecules: oxygen, nitrogen, carbon dioxide, hydrogen and so on."

Caroline: "Yes, I remember learning that last year from my eighth-grade teacher."

Mary continued: "Pressure on a surface is the result of collisions between molecules and the surface. The more collisions per second, the greater the pressure. What do you suppose would happen to the pressure if the concentration of molecules were increased?"

Caroline: "The number of collisions per second would increase?"

Don: "You sound uncertain."

Caroline: "If there are more molecules near the surface, there will be more collisions in the same amount of time, right?"

Don: "Yes, the greater the concentration of molecules the greater the pressure—what else determines the pressure that a gas exerts on a surface?"

Caroline: "How hard the molecules hit the surface?"

Don: "Yes, and what determines how hard they hit the surface?"

Caroline: "Their speed."

Don: "Yes, and their speed is determined by their energy, and their energy is determined by temperature."

Caroline: "It is?"

Don: "Think of water as it's being brought to a boil. As it gets hotter it becomes increasingly agitated, and the more agitated it becomes the faster the water molecules move around."

Caroline: "So air *pressure is determined by the concentration of molecules multiplied by temperature*?"

Don: "It is. And that simple relation you just derived, between the pressure that a gas exerts and the concentration and temperature of the gas, is called the *ideal gas law*. It's an excellent approximation for dry air at sea level and ordinary temperatures. Since the concentration and temperature of air beneath the table is the same as the concentration and temperature on the top, the pressures balance and the table is unaffected."

Fluids Can Be Made to Defy Gravity by Manipulating Pressure

Don knew that Caroline had learned how to calculate the area of a circle, but he was curious about her ability to apply what she had learned.

"If the glass is a circular cylinder with a diameter of three inches, how much atmospheric weight would be on the surface of the milk?"

Caroline thinks a bit, then says: "The area is πr^2 and r = 1.5 inches and the weight is 14.7 pounds per square inch—so it's 3.14 × 1.5 × 1.5 × 14.7 pounds. Can I borrow your phone?"

Her father hands her his phone.

Caroline used a calculator that her father had downloaded to find the answer: 103.8 pounds. "That's incredible; it's more than I weigh. This is really neat—I can figure out stuff that I could never have even guessed if I didn't do the calculations."

"Well if you think that's neat then you'll love science: there are so many startling discoveries that have been made just by calculating."

Caroline then decided to calculate the atmospheric weight on the entire table and was shocked when she found that the weight of the atmosphere on the table is nearly 59,000 pounds: "That can't be right: why wouldn't the table collapse?"

Mary explained that the air pressure is the same on all parts of the table—the air on the bottom of the table pushes up with the same pressure that the air on the top of the table pushes down.

Caroline: "I'm confused. Isn't the weight on the top of the table the result of all the air from the tabletop to the top of the atmosphere?"

She then continued, without waiting for an answer. "The air on the bottom of the table only extends to the floor. How can it be pushing up with the same force?"

Mary: "When you say from the tabletop to the top of the atmosphere, it sounds like you're thinking of air as a continuous substance like a fluid. There's a better way to think about it

."Remember that air is composed of molecules: oxygen, nitrogen, carbon dioxide, hydrogen and so on."

Caroline: "Yes, I remember learning that last year from my eighth-grade teacher."

Mary continued: "Pressure on a surface is the result of collisions between molecules and the surface. The more collisions per second, the greater the pressure. What do you suppose would happen to the pressure if the concentration of molecules were increased?"

Caroline: "The number of collisions per second would increase?"

Don: "You sound uncertain."

Caroline: "If there are more molecules near the surface, there will be more collisions in the same amount of time, right?"

Don: "Yes, the greater the concentration of molecules the greater the pressure—what else determines the pressure that a gas exerts on a surface?"

Caroline: "How hard the molecules hit the surface?"

Don: "Yes, and what determines how hard they hit the surface?"

Caroline: "Their speed."

Don: "Yes, and their speed is determined by their energy, and their energy is determined by temperature."

Caroline: "It is?"

Don: "Think of water as it's being brought to a boil. As it gets hotter it becomes increasingly agitated, and the more agitated it becomes the faster the water molecules move around."

Caroline: "So air *pressure is determined by the concentration of molecules multiplied by temperature*?"

Don: "It is. And that simple relation you just derived, between the pressure that a gas exerts and the concentration and temperature of the gas, is called the *ideal gas law*. It's an excellent approximation for dry air at sea level and ordinary temperatures. Since the concentration and temperature of air beneath the table is the same as the concentration and temperature on the top, the pressures balance and the table is unaffected."

Caroline: "That's pretty simple, but I wanted to ask—is the physics that underlies the barometer similar to the physics that underlies the gravity defying fluid in a straw?"

Don: "What do you mean by the gravity defying fluid in a straw?"

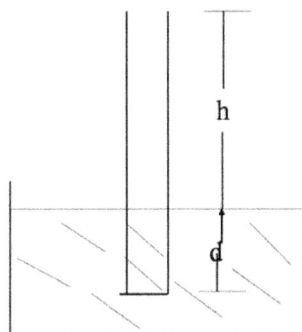

Figure 2. A straw partially submerged to an arbitrary depth, d, is placed in a liquid and then removed after its open end is covered. The air between the liquid's surface and its closed end is at atmospheric pressure and remains so after the covered straw is removed.

Caroline: "What I mean is that when I place a straw in a glass of soda, or any fluid, and put one of my fingers over its opening before removing it, then when I do remove the straw the fluid doesn't fall out but just remains there, as though defying gravity."

Don: "Yes, it's a problem in a branch of physics called statics, and understanding situations such as a fluid defying gravity is just a matter of identifying the forces and their magnitudes acting on the soda or water or whatever—but sometimes that's easier said than done. When you place the straw in your soda, the air above it will be at atmospheric pressure, and even after you cover the straw the air inside it is still at atmospheric pressure. It is evident that so long as the straw is submerged and uncovered, the external pressure at the base is equal to the internal pressure at the base—they're both equal to atmospheric pressure plus the pressure of the water at depth d. But when you remove the straw and keep the top covered, the situation changes: at first the pressure inside at the base is slightly greater than outside, so some liquid begins to leave the straw. As it does, the trapped air expands and its pressure decreases, until the inside and outside pressures balance."

Caroline: "So what's the answer?"

Don: "You're not going to get off the hook that easily—if you think about it, you'll be able to figure it out."

Caroline (sounding doubtful): "I will?"

Mary (lightheartedly): "Your father's being a bit of a magician and somewhat misleading. In fact, soda, or whatever fluid, does leave the straw: it has to because the internal pressure driving it out is greater than the outside pressure keeping it in, but the amount that leaves before the pressures equalize is so small that it's very difficult to notice."

Caroline: "OK, finally an explanation that makes sense—at least it's not magic, but it's also not convincing."

Mary: "What do you suppose happens next, what happens when the fluid starts to leave the straw?"

Caroline didn't know what to say, but then after some thought realized the obvious: "When the amount of fluid inside the straw decreases, the internal pressure must also decrease, and maybe the inside and outside pressures eventually equalize, but that doesn't sound very compelling; wouldn't all the fluid have to leave the straw before the pressures equalize?"

Mary: "What about the air above the fluid, what happens to it—does it remain at atmospheric pressure?"

Caroline: "Oh, I think I get it—since the volume of the straw occupied by water has decreased, and the total volume of the straw of course doesn't change, there's more room for the air molecules above the fluid to move around, and as a result their contribution to the total pressure decreases."

Mary: "Yes, that's the key: the air pressure above the straw drops below atmospheric pressure."

Then Caroline took out her notebook as she acknowledged that it was beginning to make more sense, but remained puzzled: "It's not obvious to me that all it takes to equalize the pressures is the loss of an amount of soda that's so small, it's not easily noticed."

Mary (clarifying): "Let's ask what the pressure difference is as soon as the straw is withdrawn from the soda. We'll denote by W and A the pressure due to water and air respectively and let r, which will turn out not to matter, be the radius of the straw. Let's suppose the glass is reasonably full, so $h \approx 0.2$ m and take d, the depth of the straw, as 0.06 m.

The density of water is $\rho w = 1000$ kg/m³ and the gravitational acceleration in SI MKS units is g = 9.8 m/s².

Then:
$$W = (\rho w g d \pi r^2) / (\pi r^2) = \rho w g d = 1000 \times 9.8 \times 0.06 = 588 \text{ Pa}$$

"As we were just saying, the fluid accounts for only part of the pressure inside. The rest comes from the air above it, initially at atmospheric pressure. So the total downward pressure inside is Patm + W, while the pressure outside is Patm.

Since: $P_{atm} = 100,000$ Pa, the fractional difference between the two is: FD = W / P_{atm} = 588 / 100,000 $\approx$ 0.0059.

"In other words, the imbalance is only about six-tenths of one percent—tiny, but enough to let a little liquid escape until things equalize."

Caroline: "It's beginning to become more understandable, but I still don't have a good intuitive feeling about the air pressure reduction and fluid loss needed to achieve equilibrium."

Mary and Don: "Caroline, you should try to figure that out for yourself."

Caroline looked a bit bewildered and her mother said, sympathetically, that she and Don understood—being asked to calculate something you've never before thought about is always a bit shocking, and the first thing you need to do is get over the shock.\

Mary reminded her daughter that the total internal pressure at the base of the straw, while it is still submerged, but after the top is sealed is $P = P_{atm} + W$, where the pressure above the fluid remains at P_{atm} + since the straw is sealed, and W is the fluid pressure. After the straw is removed, the pressure differential forces some water to leave the straw and the depth of water column changes from d to d - δ. Since $\pi \delta r^2$ is the volume of fluid that must leave the straw to bring the internal and external pressures into equilibrium, the goal is to find δ.

Caroline was bewildered and decided to get her mind off the problem by going to her room and practicing the violin. She returned to the discussion a few days later, notebook in hand, containing a derivation that she hoped was correct.

Caroline (with enthusiastic excitement in her voice): "Mom, Dad, I derived and solved a quadratic equation for δ, but the solution was very complicated. When I told my friend Noreen about it, she showed me a neat way to simplify the result, and the simplified expression was still very accurate. It turns out that

$$\delta \approx h \times (W / P_{atm})$$

and after substituting numbers I found $\delta \approx 0.12$ cm.

"In other words, the small fractional imbalance we calculated earlier—$WP_{atm} \approx 0.006$—is exactly the ratio that shows up here. The reduction in the height of the liquid is proportional to its initial height multiplied by that fraction. So if the straw is immersed to a depth of 6 cm, the percent of liquid lost before equilibrium is established is tiny: $100 \times 0.12 / 6 \approx 2\%$."

Caroline of course wanted to know the origin of that magical relation—or at least it looked magical—that Noreen showed her somewhat hurriedly but didn't explain. Her mother responded by giving Caroline relevant readings on something called the binomial theorem and told her that she or her father would help if she had difficulty understanding any of it. Caroline acquiesced, but then in an afterthought lamented: "This is all very interesting, and I think I now understand pressure, but you haven't told me anything about the sphygmomanometer."

Mary: "I actually haven't thought much about it, and doubt that I can explain it easily without doing some research. But I'll tell you someone who will know right away—your Aunt Zoe. As you know she's a professor of biology and she'll be able to explain it better than I can—and I'm sure she'd love to speak with you; she hasn't heard from you in weeks."

Atmospheric Pressure is 6 - 8 Fold Greater Than the Arterial Pressure Exerted by Blood in a Healthy Adult

The next evening after dinner Caroline phoned her aunt.

"Hi Aunt Zoe, this is Caroline. Mom said you can help me understand blood pressure.

Zoe: "Sure, I can try. But how did you become interested?"

Caroline: "Because Mom said her blood pressure is high"

Zoe: "Oh, I'm sorry to hear that. Well, high blood pressure runs in our family, so it's not surprising."

Caroline: "It runs in our family. What does that mean"

Zoe: "For me and your mother, it means both our parents had high blood pressure."

Caroline: "How is it that you and Mom tend to have high blood pressure? Do Dad's parents also have high blood pressure?"

Zoe: "No, I don't think so. And not necessarily everyone in our family does. But the explanation is another story, Caroline, and a little complicated—can we discuss it another time?"

Caroline: "OK, Dad is compiling what he calls a rainy Saturday list, which has all the questions I ask, which would be too distracting to answer as he tries to explain pressure or anything else—I'll ask him to add it to the list."

Zoe: "Great. Blood pressure might be easiest to understand if I describe the procedure for measuring it. I know your mother has a device for measuring blood pressure at home."

Caroline: "Oh, you mean the sphygmomanometer? She keeps it in her office."

Zoe: "Wow, you know what it's called—and you pronounce it so well. It's a tough word—sphygmo comes from ancient Greek and is roughly translated as pulse, or throb; manometer is Greek/Latin and is roughly translated as measuring device. After we talk, ask your mother if you can use the one she has in her office; trying it out will help you remember what I'm going to say.

"Blood pressure is normally measured by first blocking the flow of blood in an artery, a tube-like vessel in your body that moves blood away from your heart. Arteries provide oxygen to all the cells in your body—you probably heard your science teacher talk about them."

Caroline: "No, I haven't learned about arteries yet."

Zoe: "That's OK, I'm sure you'll still be able to understand what I'm about to say. The first step is stopping the flow of blood in an artery. For example, the flow of blood in the artery of your upper arm can be blocked by inserting your arm into the cuff of the sphygmomanometer. The cuff is expandable, and when it's inflated it places pressure on your arm and the flow of blood is blocked—we say the artery is occluded."

Caroline interrupts: "How do you spell occluded?"

Zoe: "O C C L U D E D."

Caroline: "I like that word."

Zoe: "It comes from the Latin word occludere, meaning to shut, or close off.

"Getting back to the explanation, although blood flow in your artery stops when the cuff is inflated, your heart continues to beat and every beat pushes blood. If you're wondering what happens when blood flow is blocked, the answer is that pressure is exerted on the cuff, which pulses slightly, and on the arterial walls, which expand a little. For most people, the occlusion is more than enough to stop blood flow, i.e., if it were very slightly less, blood still wouldn't flow, *but as pressure is slowly released the beat of your heart will start to drive blood through the artery. The pressure at which blood just begins to flow is called the systolic pressure, which is the pressure exerted by blood on your artery when your heart beats, i.e., when it contracts. The diastolic number is the pressure between heart beats when blood flow returns to its unimpeded value. The two pressures have units of mm, and are written as systolic over diastolic, e.g., 130/76.*

"I believe your father already taught you why mm can be used as a unit of pressure, and that you know that atmospheric pressure under standard conditions is 760 mm."

Caroline: "Yes, he did, so I guess that if the diastolic reading were 76 it would mean the pressure is 10% that of the atmosphere, when the atmospheric pressure is measured at sea level, in dry air that is at a temperature of 32 degrees Fahrenheit?"

Zoe: "Wow! Where did you learn all that?"

Caroline: "I read about it in a physics book after Dad told me about the metric system. I found out that in most countries of the world, temperature is measured *in degrees Celsius, in which 0 degrees is used as the freezing point of water, and 100 degrees is defined as the boiling point*. It seems simpler than using 32 degrees for freezing and 212 degrees for boiling."

Zoe: "It is. And you took the words right out of my mouth—I was just about to say that *a blood pressure reading of 76 means the pressure of blood flowing through the artery in your arm can sustain a column of mercury 76 mm high, or about 3 inches."*

Caroline: "Thank you, Aunt Zoe. I think it's incredible that the pressure inside a tiny artery can be measured so accurately."

Zoe: "You're welcome, Caroline. If you'd like to learn more, I'll send you some references. And by the way, the measurements, unfortunately, are not very accurate, and they also depend on whether you're sitting, lying or standing, so you need to be careful to compare different readings under the same conditions. If your mother took her blood pressure when she was standing, and the nurse usually takes it when she's sitting, that could account for why it seemed high."

Caroline: "Why is that?"

Zoe: "Let's save the answer for another time. Ask your father to add it to his rainy day list."

Caroline turned to her mother: "That was great—but Mom, I still want to know about the stock market."

Mary hugged Caroline goodnight: "It's time for bed, my lovely, inquisitive and exhausting daughter—can we save that for another day? Say goodnight to your father; he's in the kitchen snacking."

Conversation 2

Timbre and the Sounds of Music

Caroline's curiosity about sound and her love of music spark conversations that hold some surprises. These include the incredible sensitivity of the mammalian auditory system; the way sound propagates through air; the reasons the same notes played on different instruments sound different; the nature of waves that don't travel; the emergence of vibrato in the human voice; and the similarities and differences between sound, water, electromagnetic, matter and gravitational waves.

Figure 1. DALL-E conception of a pre-teen in her room enjoying her violin

Caroline's Home is Steeped in Beautiful Noise

The signs and sounds of music permeate the Angstrom household. Don Angstrom, Caroline's father, is a renowned pianist—his skills may not have achieved global fame, but they are undeniably exceptional. Like most musicians, Mr. Angstrom is a multi-instrumentalist. Various rooms are adorned with speaker systems, digital audio workstations, a couple of trombones, a harmonica, a violin, two saxophones, and even an erhu he acquired nearly two decades ago during a post-graduation visit to Beijing. And taking pride of place in the living room stands the grand piano—a magnificent 10-foot Fazioli, a gift from Mary to her husband on their fifth anniversary. Mary herself is an accomplished soprano, often serenading a Schubert Lieder as Don provides accompaniment, while Caroline is a budding violinist.

Some of Caroline's earliest memories were of her mother singing. Mary sang primarily for enjoyment, but occasionally practiced with intent, modulating the sounds of specific notes. Caroline was fascinated listening to her mother produce even simple sounds like vowels—inhaling deeply into her diaphragm, then letting those beautiful high notes flow effortlessly: AAAA... EEEE... IIII... OOOO... UUUU... holding each for what seemed an eternity.

Caroline still remembered how surprised she was when her mother explained that she was singing each of the vowels on the same note. She had just learned about pitch and tone, and thought that if the note was the same, the sounds should also be recognizably similar regardless of which letter was sung. Furthermore, why did the same note sung on the same vowel by different sopranos sound different, or the same note played on two different instruments? It was, in the immortal words of Oscar Hammerstein's King of Siam, "a puzzlement."

Her mother said she understood why Caroline was surprised and indicated that the explanation involved something called timbre. She added that understanding timbre required some understanding of the connection between music and the physics of sound. Mary was about to begin an explanation when her husband walked in—and without missing a beat she suggested they ask him for an explanation. He was, in fact, just the right person because in addition to being a renowned musician, he had majored in physics in college.

Caroline's father, although a distinguished musician, had struggled for years to choose between a career as a physicist and a career as a musician, before realizing he'd never be satisfied pursuing music only as an avocation. Caroline recalled how her mother would sometimes tease him about his career choice. "Don," she would say, "I think you just enjoy taking bows—can you picture a physicist taking repeated bows after a lecture? Or even a single bow?" Don would laugh and reply, "And when does an audience ever shout 'Encore!' after a scientific talk? But performers…"

Caroline still cherished the memory of their playful banter, enjoying their light humor and the obvious delight they took in one another. She also remembered that particular exchange because she was surprised when her mother, who was rarely at a loss for words, struggled to answer her question: What is a physicist? After a moment of frustration, Mary blurted out, "Physics is a search for understanding—understanding everything on Earth and in the entire Universe. Yep, all the processes that occur here and beyond are grist for the physicist's mill."

Though Caroline, about to start ninth grade, didn't fully grasp the explanation or the almost sarcastic tone when her mother said *entire Universe*—she sensed there was more to it. Mary, no stranger to the sciences herself, had majored in aerospace engineering at Boston University before pursuing graduate studies in mathematical finance at MIT. She often found herself mildly disdainful of the arrogance she noticed in physicists and mathematicians who believed they were revolutionizing Wall Street, just as they had transformed the biomedical sciences. But she let that discussion drop for now. Caroline couldn't be expected to understand all the nuance and ambiguity of conversation, but she hoped her daughter's probing, insight, and curiosity would grow steadily and bring her lasting satisfaction.

Caroline progressed steadily and at a reasonable rate, pretty much as her mother had hoped. Now a high school sophomore who had completed freshman physics, she was curious about almost everything and took special delight in trying to understand the everyday occurrences that most people take for granted.

One Saturday afternoon as Mary was completing practice for a *Saengerfest* that her recently formed choral group had organized for the coming evening, Don casually asked his daughter if she remembered wondering why the same note generated by different sources sounds so different.

"How could I forget—Mom still practices singing several days a week, and now she often modulates even the length and loudness of the notes." Then her father said something puzzling: not only did his wife sing different vowels on the same note, she sometimes sang base words such as *bat*, *love*, *sing*—or even affixes such as *un* and *re*, on the same note.

Caroline frowned. "Wait, hold on. What's an affix? And what's a Saengerfest?"

The shift in conversation had piqued her curiosity and her father, sensing an opportunity to explore a new concept with her, leaned in. "As you know, *words can be built from smaller components, but what you might not know is that the way we construct and modify them follows patterns, just as music does.*"

Caroline: "I'm listening."

Don, seeing that he caught Caroline's attention with something other than classical science, was eager to elaborate a bit more and then let the conversation take its course.

"With respect to word *affixes, they're often just a few letters and a single syllable which has no meaning of its own, but can modify the meaning of a word* to which it's attached. For example, the prefix *un*, which I'm sure you heard your mother sing, when attached to words such as *recognizable* and *able* transforms them into their opposites: *unrecognizable* and *unable*. However, not all words that can't be decomposed into smaller meaningful units are affixes. For example, the word *bat* has a meaning: it can be a piece of wood designed to hit a ball, or it can be an animal. But in both cases it can stand alone, so it's not an affix. Stand-alone words that can't be divided into smaller meaningful units are called *free morphemes*, as opposed to *affixes*, which are *bound morphemes* that need to be attached to free morphemes to be meaningful.

"Some languages such as German have lots of long compound words that are created by stringing free morphemes together. When you look up the meaning of *Saengerfest*, you might want to delve a bit into that interesting aspect of German. By the way, the word *morpheme* is derivative of *morphology*, which is the branch of linguistics that studies the evolution and development of vocabulary."

Caroline: "Sounds interesting, no pun intended."

Don: "It is, it's very interesting. At some point you'll no doubt start asking questions about linguistics, and it wouldn't surprise me at all if you find it to be as fascinating a subject as physics and mathematics."

Caroline: "Is the relationship between *morpheme* and *morphology* similar to the relationship between *molecules* and *chemistry*? I mean *morphology* is a discipline and the fundamental unit of that discipline is the *morpheme*, just as *chemistry* is a discipline, and its fundamental unit is the *molecule*. They're similes, or perhaps they're metaphors for one another—in either case I would say that morphemes are the molecules of morphology."

Don, surprised by his daughter's thoughtfulness, replies with astonished hesitancy: "Actually I never thought of that, but it sounds like a reasonable analogy. And where did you learn about metaphors and similes? You must have a great English teacher."

Caroline, always a bit too impatient: "I do, I have an excellent English teacher, but we're digressing, and I'd like to get back to music because I've made plans with some of my friends, and I don't have much time before I'm supposed to meet them."

Don: "Wow! It seems that you've got more going on in your life than I have in mine."

Caroline responds in the somewhat self-centered and almost disrespectful way teenagers sometimes act: "I do—so let's get on with it."

"I've now had a year of physics but I still don't understand why, when Mom sings different vowels on the same note, they don't sound as similar as I would have expected. I thought of asking you again for an explanation, but for some reason I didn't."

Don, worried that something he said might have discouraged Caroline from asking the question: "Caroline, please don't hesitate to ask questions; no one will think less of you for not knowing something. In addition—and this is important—people usually welcome questions. Questions show that you're interested in understanding what they're saying that you're paying attention."

Caroline: "OK, from now on I'll gather up my courage."

Don: "Did I ever tell you about the boy who kept asking his father questions? The father couldn't answer any of them, and the son finally asked: 'Dad, am I annoying you by asking so many questions?' The father replied, 'Not at all—how else are you going to learn?'"

Caroline: "Oh Dad, really: how corny. Now, can we please get to the explanation?"

Don: "Nothing like a corny joke to lighten a serious conversation. OK, let's return to your question. But before I try to answer it, we need to go back to basics."

Waves Permeate the Universe

Don began with a very broad perspective, indicating that sound is one of several different kinds of waves including electromagnetic, gravitational, and matter waves—though he mentioned them with some trepidation because he was sure Caroline had never heard of any of them. He cited *light as an example of an electromagnetic wave* and briefly mentioned gravitational and matter waves, noting that they involve deep concepts in general relativity and quantum mechanics. As he expected, Caroline was disappointed by her father's lack of willingness to elaborate further.

Caroline: "Dad, I'd almost prefer that you say nothing at all rather than use words that mean nothing to me."

Don: "I understand your frustration but I did give you some information—you now know that waves in addition to sound and light exist and that they play central roles in areas of physics called quantum mechanics and general relativity. I can also give you some introductory material written for a lay audience if you'd like to do some reading, but the subject is complicated and if we digress too far we'll never get to your main question."

All Waves Transport Energy and Have Common and Unique Properties

Don went on, "I did want to make a few important points: one is that *all waves, no matter what type, earn their living in the same way—by transporting energy*—but they do so in different ways."

Caroline: "Such as?"

Don: "I'll give you two examples in a little while. For now, I'll just note that in addition to transporting energy, *all waves exhibit a number of common properties*. And here again, I'll use some words you may never have heard of but won't elaborate on them because you can easily dig deeper on your own—and we do need to move on. The properties I have in mind *are reflection, refraction, dispersion, and diffraction*—all of which you will soon encounter in depth. Finally, and I know you've come across this in your freshman physics class: the length of a wave, for example, its peak-to-peak distance, multiplied by its frequency, is equal to its speed."

Caroline: "Yes, that one I understand, and I'm actually glad you're not going to elaborate because I am eager to move on."

A Musical Note Has a Complex Structure

Don explained that sound is a pressure wave that propagates through air, liquids, and solids. "Let's start with something simple: a tuning fork," he said, holding one up. *"When you strike a tuning fork, the prongs vibrate and create a nearly pure tone—a sound with a single frequency. The purity comes from the fork's design and its high-quality steel, which minimizes additional vibrations."*

Caroline: "Are you implying that only a tuning fork will produce a pure tone? If I play a single note on my violin, doesn't the string vibrate at only the frequency of that note?

Don: "Actually, no it doesn't it. In fact musical instruments produce complex vibrations, and that complexity is key to understanding the unique character of different instruments. We can talk more about that later; for now let's stick to the tuning fork which is easier to understand.

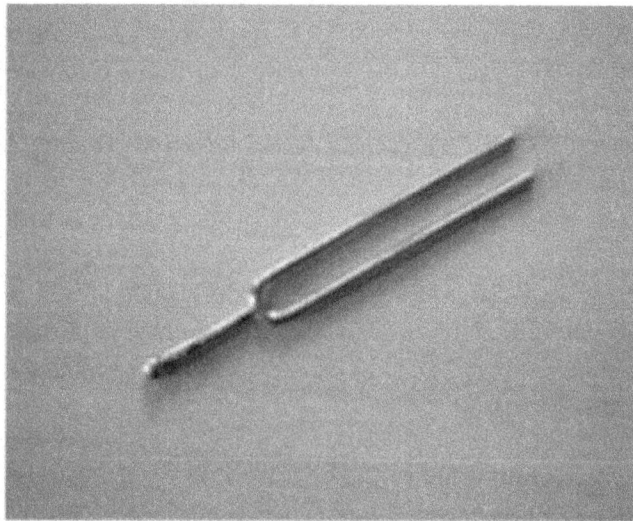

Figure 2. A typical tuning fork. Rotation by 180 degrees about an imaginary line that passes through its stem and runs parallel to the prongs leaves the figure invariant. Because of this symmetry, the prongs vibrate inward and outward synchronously, i.e., they move inward together and stop simultaneously when they're at their closest distance, and then outward together, stopping simultaneously when they're furthest apart.

Tuning Forks Generate Only the Fundamental Frequency of a Note

Don's comment about the complexity of a note had caught Caroline's attention, and she listened closely as he continued: "When a fork strikes a surface, the prongs vibrate at a frequency determined by their length — the longer the prongs, or tines as they're sometimes called, the greater their mass, the slower they vibrate, and the lower the pitch. In fact, the *frequency varies inversely as the prong length squared*."

Caroline: "That reminds me of gravity, which also follows an inverse-square law."

Don: "It's true that gravity weakens with the square of the distance, but it's important to understand that the relationships arise in very different contexts. The tuning fork's frequency depends on the physical properties of the prongs—their length and mass—which are unrelated to gravitational forces."

Caroline: "It's surprising that so many things follow the same kind of mathematical relationship."

Don: "It is, but it's more a reflection of how nature's patterns often lead to similar mathematical descriptions rather than a direct connection between the underlying physics. Just to follow up on the length-pitch relationship, can you tell me the fundamental frequency of a fork whose prongs are twice as long as those that produce middle C?"

Caroline: "Well, if the frequency varies inversely as length squared, I guess it would be four times lower."

Caroline quickly estimates: "262/4—around 65 cycles per second?"

Don: "Yes, that's a good estimate. And, of course, that's a C, but two octaves below middle C.

"By the way, if you do some reading, you'll usually see the word *hertz*—abbreviated Hz—used instead of 'vibrations' or 'cycles' per second. *A hertz is one vibration per second* and honors the German physicist Heinrich Hertz, who contributed greatly to our understanding of electromagnetic waves."

Her father then adds with a sly smile, "Although Hertz might never have bowed after delivering a scholarly lecture, his name is probably spoken and written tens of thousands of times every year all over the world, and that might well remain true for hundreds of years into the future."

Don looks at his daughter, as though anticipating a question, and then says, "I know, now you're going to ask me what an electromagnetic wave is. Let's wait a while before discussing that. The concepts are deep, I'm not an expert, and I'll need some time to organize my thoughts."

Caroline: "OK—and you're right, that's exactly what I was going to ask."

Don: "Getting back to what I was saying, when the fork strikes a surface, the prongs move in a repetitive fashion—inward together, then outward together, then inward together, etc. What do you think a plot of the distance between prongs versus time would look like?"

Caroline: "It would increase, then decrease, then increase. I guess it's a wave."

Don: "It's a particular type of wave, a very regular wave—you studied it in algebra."

Caroline: "A sine wave?"

Don: "Yes, the motion is typically represented by a sinusoidal curve—*a sine or a cosine* (Figure 3). They're exactly the same, only shifted from one another by π/2 radians, or 90°. Every position along each prong moves back and forth in sinusoidal motion, with the greatest amplitude of vibration at the ends of the prongs, and essentially no motion at all where they join. The result is a series of compressions and rarefactions of the surrounding air.

"I'll say more about such patterns of motion when we discuss the vibration of piano strings, but for now, here's the connection with sound: as the prongs move away from one another, they compress air molecules, creating a region of high pressure.

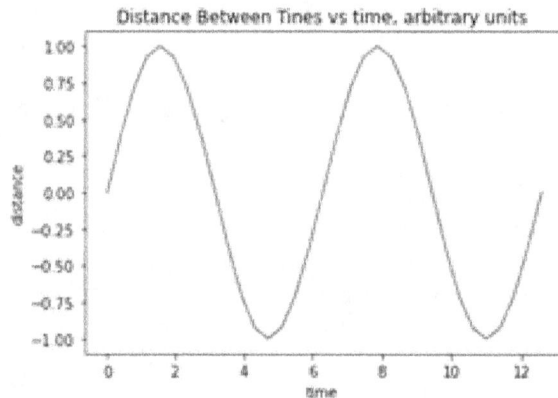

Figure 3. When the fork is struck at time 0, the prongs move from their resting position, achieve maximum (or minimum) separation, which in this example occurs at approximately 1.6 units of time, and then move together achieving a minimum separation at approximately 4.7 units of time. A notion of scale can be obtained by noting that if the prongs are designed to produce middle C, a complete cycle, which on this figure occurs at approximately 6.3 time units, would take approximately 1/262 = 0.0038 seconds. The variation in distance between prongs is directly proportional to the variation in energy propagating away from the tuning fork, i.e. the ordinate also describes pressure to within a multiplicative constant.

"As they move toward one another, the air rarefies, creating a region of low pressure. Their vibrations are thereby converted into waves of energy that move through the air and can be detected by the human ear—or, for that matter, by any mammalian ear that encounters them."

To help Caroline visualize what he's saying, he shows her a representation of the compressions and rarefactions of air molecules as sound energy propagates from source to detector (Figure 4).

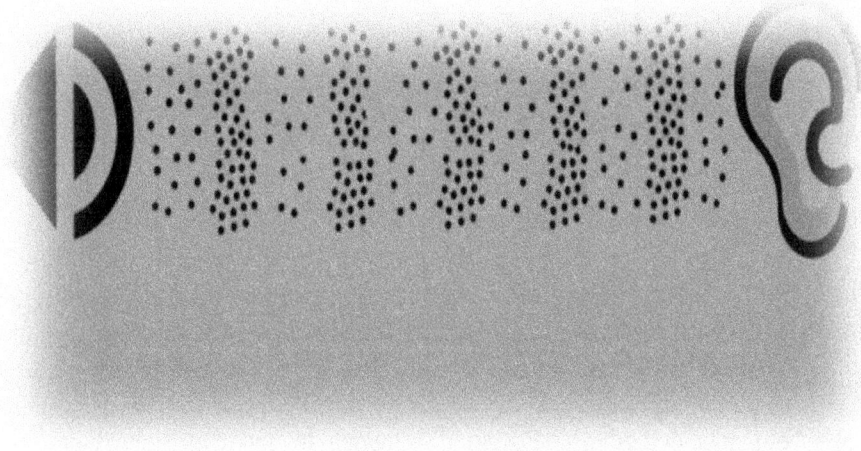

Figure 4. Cartoon of compressions (columns of dense dots representing air molecules) and rarefactions (sparse dots) that transmit energy from source (speaker) to detector (ear). The pressure, which varies from high to low continuously, is often represented by a sinusoidal wave as in Figure 3.

Sound Moves Forward but the Molecules of Air Through Which it Moves Only Vibrate

Caroline: "I'm a bit puzzled. The diagram seems to suggest that waves of air are moving away from the tuning fork, but if sound is carried by moving air, wouldn't you be able to detect a breeze, especially for loud sounds?"

Don: "I can see how this could be confusing. If sound were caused by pressure waves of moving air, that would in fact create a minuscule breeze. But sound can move through air that is perfectly still."

Caroline: "Then how does the sound propagate?"

Don: "Sound is essentially the propagation of energy. The easiest way to understand it is to first consider what still air is like at a molecular level in the absence of sound. Picture molecules moving around randomly in two adjacent regions of air that have the same temperature, volume, and pressure. Molecules in each region collide with one another, but there's no net transfer of kinetic energy between regions because there's no difference in the number of molecules in the two regions, and no difference in their kinetic energies."

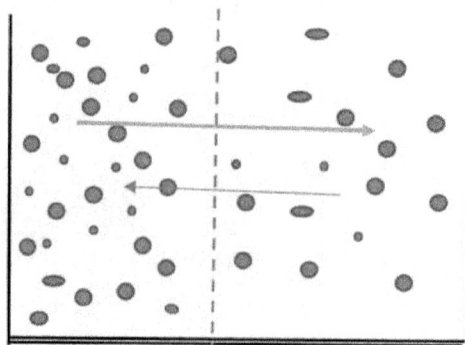

Figure 5. The cartoon illustrates the idea that the rate at which molecules move from the higher concentration region (left of the dashed line) to the lower concentration region is greater than from lower to higher. As a result there will be a net transfer of energy from left to right, and therefore a net energy flow in that direction. Although energy propagates, individual air molecules oscillate around their equilibrium positions, as described in the text.

Don: "When a tuning fork generates waves of compressed and rarefied air, however, adjacent regions have different numbers of molecules (Figure 5). Molecules in the high-concentration region will cross into the low-concentration region where they transfer energy by collision. Similarly, molecules in the low-concentration region will collide with molecules in the high-concentration region. However, the number of crossings from high to low concentration will be greater than the number from low to high. Consequently, there will be a net transfer of energy, which is the propagation of sound.

"When the prongs reverse direction, the region that was compressed will become rarefied, pulling molecules back so that they are essentially where they were when the cycle started. The result is an energy wave that propagates longitudinally, i.e., in a direction more or less perpendicular to the long axis of the tuning fork—but air doesn't move with it."

Caroline: "I think I get it: the vibrating prongs cause air molecules to vibrate around their original positions while the kinetic energy of those molecules is transferred in such a way that it flows outward from the source as an energy wave that can be detected by our ears."

Don: "Yes, air waves of sufficient pressure can be detected by our ears—or in fact they can be detected by the ears of any mammal, though the required amount of pressure will differ for different species. We'll talk more about waves when we discuss the piano, but for now just know *that a tuning fork will vibrate almost entirely at the fundamental of the note it's designed to generate, with the result that you'll hear a nearly pure tone.*"

Frequency, Wavelength, and Speed Are Connected

Don: "Now that we understand how *energy moves through air without the air itself traveling*, let's explore the relationship between the length, frequency, and speed of a wave.

"Because the shape of a sound wave—the pattern of compressions and rarefactions of air produced by a pure tone—is sinusoidal, it can be characterized by just two numbers: one for its maximum amplitude (the height of the peak), and the other for either its frequency (the number of peaks per unit time) or its wavelength (the distance between two corresponding points on successive waves). Frequency and wavelength are easily related: the length of a wave is its speed divided by its frequency."

Don then drew a wave and asked Caroline what the distance between peaks would be if the frequency were 440 cycles per second, the standard pitch to which musical instruments are tuned.

Caroline quickly grabbed her calculator, googled "speed of sound," and the screen instantaneously flashed *1,125 feet per second* in dry air at 20°C. A flood of thoughts occurred nearly simultaneously. She understood why the result was qualified, recalling from her previous discussions that pressure depends on temperature. She was also thinking, "1,125 ft per second at sea level," even though the sea level qualification wasn't explicitly included. She was pleased with herself for adding to what she found, even as she wondered how to find the distance between peaks—or between any corresponding points in consecutive cycles—in terms of speed and frequency.

Then, she suddenly understood why speed equals frequency times wavelength. She turned to her father and blurted out excitedly that if she imagined herself standing still and counting the number of peaks that passed her each second, the wavelength would be the speed of the peaks divided by the number of peaks that pass her (Figure 6). *"If sound travels 1,125 ft in one second, and I see 440 waves go by in that second, then the distance covered by a single cycle—or equivalently, the distance between corresponding points on adjacent waves—must be 1,125 / 440 = 2.6 ft."*

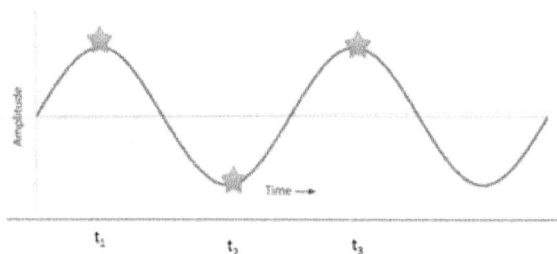

Figure 6. A sinusoidal traveling wave, illustrating how amplitude changes with time, as measured for example in seconds. The reference point, indicated by the star, is at its maximum value at time t_1 seconds after the start of the wave, at its minimum value at t_2 seconds, and again at its maximum after t_3 seconds. It, therefore, completes 1 cycle in time $t_3 - t_1$ seconds, and has a frequency of $1/(t_3 - t_1)$ cycles per second. If $t_3 - t_1$ is 4 seconds, the wave's frequency is 0.25 cycles per second. If it's traveling at 5 feet per second, the distance between peaks is 20 feet.

Sound Propagation in Water is More Complex Than it is in Air

"That's very good, Caroline," her father says with pride, then adds, "These waves transfer energy from one location to another, which is why they are called traveling waves. Can you think of other examples of traveling waves?"

Caroline: "Ocean waves?"

Don: "Yes! Just as sound waves transfer energy through air—which remains largely stationary when there's no wind—ocean waves transfer energy through water. But while they share some similarities, there are also key differences between sound waves and water waves."

Caroline: "Like what?"

Don: "One big difference is the way molecules move to transfer energy. In *sound waves, air molecules vibrate longitudinally, back and forth along the direction in which the sound is moving, creating alternating regions of compression and rarefaction. By contrast, water molecules in ocean waves follow looping paths as the wave passes.*"

Caroline: "Do you mean the water molecules move in circles?"

Don: "Almost! In deep water, where the depth is much greater than the wavelength, the molecules move in nearly circular orbits. But as the water gets shallower, these orbits become more and more elliptical until, near the shore, the motion is mostly back-and-forth."

Caroline: "So if waves get taller, does that make the orbits more elliptical too?"

Don: "Not quite. The shape of the orbits is mostly determined by the water depth. In deep water, where waves have room to move freely, the paths are nearly circular. In shallow water, the wave base interacts with the seafloor, which flattens the motion into ellipses."

Caroline: "So the deeper you go, the smaller the motion?"

Don: "Exactly. The motion of water molecules decreases with depth, dying out almost completely at a depth of about half the wavelength."

Caroline: "What about really big waves?"

Don: "When waves get very tall compared to their length, they become steep and start forming a shape called a *trochoid*. That means the orbits of water molecules are distorted—rather than perfect circles or ellipses, the motion is more complex. Engineers and mathematicians have studied this shape in detail, but that's getting into advanced wave physics."

Don was thinking about trying to say something more about trochoids but realized that he was at the edge of his knowledge. Instead, he wisely decided not to go further, saying only that engineers and mathematicians study in detail the forces that generate trochoid formation, and that she would need a year or two of college physics before she could take a *deep dive*.

Caroline: "Why do you find puns so irresistible?"

Don: "Why do you find asking why so irresistible?"

The Human Auditory System Can Detect Pressures a Million Times Smaller Than an Ant Exerts on a Hard Surface

Caroline: "OK, let's shift the conversation a bit—from the propagation of sound to its detection."

Don: "Sure, and I'll start with something startling: *the human ear is so sensitive that it can detect pressure changes as small as three billionths of a pound per square inch.*"

Caroline: "I understand that the number is incredibly small, but I don't have a feeling for just how small that is."

Don: "Remember, the definition of pressure is force per unit area. If we express it in units used in the U.S., the human auditory system can detect a pressure variation as small as 3×10^{-9} pounds per square inch. That's roughly a million times smaller than the pressure an ant would exert on a hard surface."

Don can see that Caroline grasps the extreme sensitivity of human hearing and that she's awe-struck by the thought that the auditory systems of all mammals are probably equally sensitive. He continues by explaining that the loudness of a sound—its amplitude, or what we call "volume" on the TV—is related to the ratio of the pressure that produces it to the threshold pressure for hearing.

"To keep the numbers manageable, the amplitude of sound is specified by the logarithm of the ratio of its pressure to the threshold pressure required for hearing. In particular, the decibel (dB), the standard descriptor of loudness, is defined as:

$$dB = 20 \log_{10} (p / p_0)$$

"where lower case p is the pressure of the sound of interest, and $p_0 = 3 \times 10^{-9}$ pounds per square inch is the threshold for hearing, i.e., when the pressure drops to p_0, sound can no longer be detected."

Don emphasizes his point with deliberate care: "*The minimum pressure a sound wave must exert to be heard is strongly dependent upon its frequency.* The ear's maximum sensitivity occurs at around 4 kHz, where sounds as low as 5 dB can be detected. However, as the frequency deviates from this 'sweet spot,' the minimum pressure required for detection increases. For instance, at the extremes of human hearing—20 Hz on the low end and 20 kHz on the high end—the threshold jumps to 70 dB or higher. This frequency dependence is rooted in the anatomy and physiology of the ear."

Don pauses to retrieve a diagram (Figure 7) that Zoe had given to Mary. The sisters, coming from different backgrounds, had many insightful conversations about cochlear mechanics which they often shared with Don. Holding up the diagram, he continues, "When sound reaches the eardrum in the outer ear, it must travel through various materials: air, the solid bones of the middle ear, and the fluid-filled cochlea. Each material resists the transfer of sound energy differently—an engineer would say they have different impedances.

"To ensure efficient energy transfer, the ear compensates for these mismatched impedances. The middle ear functions like a lever system, amplifying the vibrations to minimize energy loss. Think of it like jumping between surfaces—if they're wildly different, you lose energy unless you adjust the force of your jump."

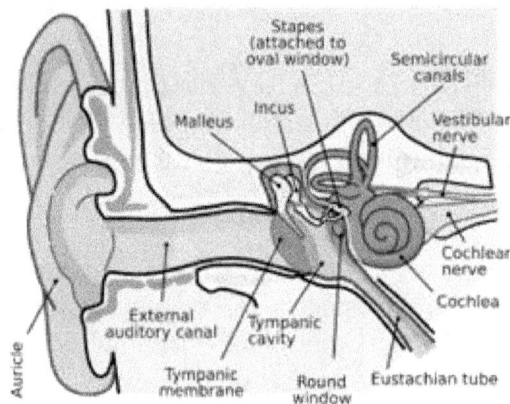

Figure 7. Sound waves enter the ear canal, causing the tympanic membrane to vibrate. The vibrations are amplified by the ossicles (malleus, incus, stapes) in the middle ear and transmitted to the cochlea where they are converted to electrical signals. The signals propagate along the vestibular auditory nerve to the brainstem, and eventually to the auditory cortex where sound is perceived.

He gestures to the figure again. "However, the system of impedance matching is optimized for mid-range frequencies (around 4 kHz, where human hearing is most sensitive). At very high or very low frequencies, energy transfer becomes less efficient due to the physical limitations of the ear's structures. This explains why sounds at the extremes of human hearing require much higher intensities to be detected—70 dB at 20 Hz or 20 kHz versus just 5 dB at 4 kHz, and about 35 dB for middle C. But remember," Don adds with a grin, "the decibel scale is logarithmic in pressure. So, Caroline," he teases, "what's the ratio of the minimum pressure required to hear a sound at 20 Hz compared to middle C?"

Caroline smirks, her expression a mix of exasperation and amusement, silently bracing for what she suspects will follow.

Caroline: "Dad, I don't know, you've thrown so many numbers at me."

Don: "Take it slowly, don't get flustered. Start with the definition of decibel. Call the decibel levels for sounds at frequencies of 20 Hz and 262 Hz (middle C) dB_1 and dB_2, and the corresponding pressures p_1 and p_2."

Caroline: "OK—for sounds with 20 Hz and 262 Hz frequencies:

$dB_1 = 20 \log_{10} (p_1/p_0)$ and $dB_2 = 20 \log_{10} (p_2/p_0)$, or

$70 = 20 \log_{10} (p_1/p_0)$ and $35 = 20 \log_{10} (p_2/p_0)$.

Now what?"

Don: "Come on, daughter, think. These are relatively simple to solve."

Caroline, after a few moments of struggling: "I think I can get the answer if I subtract one equation from the other:

$35 = 20 \log_{10} (p_1/p_0) - 20 \log_{10} (p_2/p_0) = 20 \log_{10} (p_1/p_2)$.

So $p_1/p_2 = 10^{35/20}$."

She then uses the calculator on her phone and finds that $10^{35/20} \approx 56$.

Don: "Yes, the minimum pressure required to hear a sound at the lowest frequency detectable (20 Hz) by the human auditory system is about 56-fold greater than the minimum pressure needed to hear middle C (262 Hz)."

Caroline: "So a 56-fold increase in pressure is required to hear a sound whose frequency is 13 times lower than that of middle C."

Caroline is now feeling more confident and wonders what the ratio would be at 4 kHz, the frequency of maximum sensitivity. She is stunned at the answer.

"Dad, the pressure required to detect sound at the very low or very high end of our frequency spectrum is 2,000 times greater than the pressure required at 4 kHz, the frequency of maximum sensitivity!"

Don: "The other, and perhaps more important, point is that *the frequencies of much of what we hear are well below the frequency at which the sensitivity of the human ear is optimum—around 4 kHz.*"

Caroline: "That's interesting. Why did evolution leave us so far from the optimum?"

Don: "Perhaps because if that weren't the case, we humans would have evolved in a sea of noise, including the sounds of our own physiological functions—blood flow, the opening and closing of heart valves, and breathing."

Caroline is awestruck but quickly recovers and asks the obvious but somewhat banal question.

Caroline: "Why is the log multiplied by 20?"

Don seems a little exasperated and advises his daughter to be more discriminating in her choice of questions. "Ask for help only after you've struggled a bit," he says with uncharacteristic impatience before changing the subject and leaving Caroline surprised and somewhat disappointed. But she did get the message and returned a few days later with an explanation.

Caroline: "The factor of 20 spreads the numbers."

Don adds, with a slight guess: "One of the pioneers probably chose it arbitrarily, and it became a convention."

Caroline, with a barely audible and mildly snide tone: "History again..." She then summarizes: "So, at the pressure threshold, the decibel level is 0—I suppose that makes sense, because by definition, you can't hear a sound at the threshold."

Her father confirms that her summary is on target, then reminds his daughter that what's moving through the air is an energy wave. "Therefore, it's the energy per unit time crossing your eardrum that's the most appropriate analogue for loudness." He then says nothing further, except to ask her to derive an equation for decibels in terms of power rather than pressure.

Caroline, perhaps not surprisingly, had been hoping her father would derive the equation so she could move on to other concepts. But she managed to discipline herself, and after some research, found that the power P of a wave is proportional to the square of the pressure, so $P = k \cdot p^2$. But she was stymied by the proportionality constant k—she had no idea how to evaluate it. When she pursued its meaning with her usual enthusiasm, she wound up even more lost, encountering terms like *auditory impedance*. She'd never felt so frustrated and forced herself to stop thinking about the derivation and tried to relax.

When she returned to the problem the next day, she had another new experience—a sudden realization: only the ratio P / P_0 mattered; the constant canceled. Consequently:

$$\mathbf{dB = 10 \log_{10} (P / P_0)}$$

"Except for a factor of 2, the equation for decibels is independent of whether the power ratio or pressure ratio is used."

Caroline now feels happily satisfied, and when she shows the results to her father, he compliments her as usual and explains that everyone, no matter how bright, encounters difficulty trying to understand scientific concepts and work through mathematical problems. But then, in what she thinks is a non-sequitur, her father asks if she has any idea how high the decibel level of a rock concert is.

Caroline: "I haven't the slightest."

Don: "It's about 120 dB."

Caroline immediately takes out her calculator and estimates the corresponding pressure.

$p/p_{ref} = 10^{120/20} = 10^6$

She also notes that in terms of power

$P/P_{ref} = 10^{120/10} = 10^{12}$

Caroline: "Dad, it's almost frightening! *The pressure on our eardrums from music at a rock concert is a million times greater than the threshold pressure for hearing*!" She then adds, with just a touch of sarcasm: "I guess the performers want to set the microphone gain high enough so that even the hard of hearing don't miss a single note."

Don smiles and suggests that a comparison with normal conversation would be equally interesting.

Caroline: "What's the decibel level of normal conversation?"

Her father replies that it's approximately 60 dB.

Caroline again does a quick calculation, discovers that $p/p_{ref} = 1060/20 = 1000$, and excitedly tells her father that the pressure on the human eardrum during a rock concert is 1000 times greater than the threshold for hearing during a normal conversation—in other words, it's a thousandfold greater than what we typically experience. She's stunned, but then has another thought about hearing thresholds.

She asks her father if he can predict how far away two people with normal hearing can be from one another and still maintain an understandable conversation.

Don: "Can I make the prediction? Yes, I can, but can you?"

Caroline: "I have no idea."

Don: "Here's a hint: what do you suppose the relationship is between sound intensity and distance from the sound source—how does the intensity vary with distance?"

Caroline thinks for a moment and then responds with a mix of timidity, confidence, and a feigned horror: "No, not that, not an inverse square law!"

Don: "Yep, it's the infamous inverse square law, although there's a caveat."

Caroline: "Which is?"

Don: "It strictly applies to point sources radiating in an idealized uniform medium. In actuality, reflections and absorption modify the decay rate. Nevertheless, it's worth thinking about how to use it to estimate the distance."

Caroline: "OK."

Although Caroline is excited by the surprise of something new, Don reflects somewhat somberly on the implications of the decibel level of a rock concert. He knows that chronic exposure to sounds above 90 dB can damage hearing, and sounds above 130 dB are painful—so he advises his daughter to avoid too many rock concerts.

Caroline: "Dad, I'm sorry, but I don't like the way this conversation is going. I enjoy rock concerts."

Don: "You can go to them, but don't overdo it."

Her father then diverts the conversation, asking if she can think of other wave-like phenomena that humans and other forms of life can detect. She appears stumped until the light suddenly dawns.

Caroline: "Yes," she says with a playful smile, "the light has dawned."

Caroline continues: "My science teacher mentioned what you had said in passing—that light is an electromagnetic wave—and like you, she also didn't elaborate. What's with that? Why won't anyone explain electromagnetic radiation?"

Her father responds by again postponing the conversation, saying that Uncle Zack, a physicist who does research at Los Alamos, will explain it to her, and suggests that in the meantime, she might try looking at some elementary textbooks.

He then makes the mistake of offhandedly mentioning that "There are other exotic types of waves as well, such as matter waves and gravitational waves. *Although all waves have similar characteristics, they also have unique properties.*"

Caroline: "Yes, I get that there are differences: sound waves can be transmitted through solids, liquids, and gases—and in air, the energy is propagated longitudinally. But I have no idea about the other types of waves, and what I don't get is why you keep mentioning things I couldn't possibly understand, like electromagnetic, gravitational, and matter waves."

Visible Light and Other Electromagnetic Waves Differ Only in Their Frequency

Don: "We'll get to electromagnetic waves, but it'll take a while. Right now, all I can tell you is that visible light, ultraviolet, infrared, radio waves, X-rays—they're all electromagnetic waves and differ only in frequency. *Visible light is just the portion of that frequency range our eyes can detect. That's the only difference. But that difference turns out to be profound because it causes them to interact differently with matter."*

He paused, then added with a chuckle, "As for gravitational waves... you're probably right—I shouldn't have mentioned them. All I really know is that they were predicted by Einstein's theory of general relativity, which describes gravity not as a force in the traditional sense, but as a curvature in the fabric of spacetime."

Caroline: "Spacetime? What's that? Do you mean space and time?"

Don: "No, it's spacetime—suggesting that space and time are somehow intertwined. I don't understand it—try speaking to your uncle about it. For now, let's continue our discussion of sound."

Caroline: "Sorry Dad, now you've piqued my curiosity. Can I try to reach Uncle Zack now? He'll be busy during the week, so this might be a good time. I'm afraid that if I don't do it now, it will slip through the cracks. Maybe he can give me a 30,000-foot view, and then we can continue later."

Don, recognizing that he's also curious: "OK, let's have a three-way. I haven't spoken to your uncle in a couple of weeks."

He dials the number, and after a few minutes of brotherly chatter, Caroline jumps in.

Caroline: "Hi Uncle Zack. Dad's been telling me about the physics of music, and about sound and other types of waves. He also inadvertently mentioned gravitational waves, but when I pushed him for an explanation, he said I should ask you."

Zack: "Sure, but I've got a meeting to prepare for, so my explanation will need to be brief and sketchy."

Caroline: "I'm sure anything you can tell me will be helpful. Maybe you can also point me to some introductory articles?"

Zack: "Of course. I'll send a few links afterward. But maybe the best place to start is with black holes, which are the remnants of very massive stars at the end of their evolutionary journey."

Caroline: "*Stars evolve?*"

Zack: "Yes—they're born, they mature, they age, and they die."

Caroline, flippantly: "So… stars are born? Do they reproduce sexually or asexually?"

Zack, smiling: "Cute. But seriously, 'birth' in this context isn't like biological birth. It's an analogy. Stars form through purely physical processes."

Caroline, mock serious: "Well, life also forms through physical and chemical processes."

Zack: "OK, smart aleck. Here we go. Stars form in enormous clouds of gas and dust called *nebulae*. If a region of a nebula becomes dense enough—maybe due to a nearby supernova shockwave—it begins collapsing under its own gravity."

Caroline: "Supernova?"

Zack: "We'll get to that in a minute. Just *think of supernovas as the most explosive events in the Universe.* Anyway, as the gas collapses inward, it heats up. If the core becomes hot enough, hydrogen atoms begin fusing into helium, releasing energy that pushes outward and halts the collapse. That's the birth of a star."

Caroline: "I remember learning that everything that's born dies—at least as far as we know. And I also learned that the Sun is powered by nuclear fusion. I think it's something like 14 billion years old, right?"

Zack: "It's *the Universe that's approximately 14 billion years old. The Sun was born relatively recently, around 4.6 billion years ago—but the lifespans of stars vary enormously.* Those that are much bigger than the Sun burn through their fuel in a few million years, while smaller stars can last tens of billions of years."

Caroline: "What happens when they run out of fuel, when nuclear fusion stops?"

Zack: "Now you're getting to the heart of the *matter*."

Caroline: "How did I know we wouldn't get through this without a pun?"

Zack, unfazed: "What happens next depends on the star's size. When a Sun-like star—say, between 0.8 and 8 times the mass of the Sun—runs out of hydrogen, the fusion pressure in the core drops. Gravity takes over, the core contracts, and gravitational energy heats things up again. That ignites hydrogen in a shell around the core."

He paused and turned the tables. "Want to guess what happens next?"

Caroline: "Does it explode?"

Zack: "Actually, it does the opposite. Its outer layers are slowly ejected into space over thousands of years. As it expands, its surface cools. So—what do you suppose happens to its color?"

Caroline: "I'll guess it shifts to red, but I don't know why."

Zack: "Exactly. *Sun-like stars become red giants.* The reason for the red color? Very briefly: the energy of electromagnetic radiation is proportional to frequency. *Red light has a lower frequency—and therefore lower energy—than blue. So, cooler objects emit more in the red.*"

Caroline: "That's not too bad. As the star cools, it radiates less energy and shifts to lower, frequencies—toward red. I just need to understand where that energy-frequency link comes from."

Zack: "That's another story. But our current story isn't finished yet."

Caroline: "I should've guessed. So the red giant isn't the final fate of our Sun?"

Zack: "Nope. *Eventually, the red giant sheds its outer layers and the core that remains becomes a white dwarf*—hot, dense, and small."

Caroline: "And then? Is that the end?"

Zack: "Not quite. Even though it no longer generates its own energy, it's still hot. And as you probably know, a hot object in contact with a colder one gives up heat—that's the second law of thermodynamics."

Caroline: "How long does that take?"

Zack: "Tens of billions of years. Maybe longer. Eventually, it becomes a black dwarf: cold, dark, dense, and mostly carbon and oxygen. But none have been observed—we don't think the Universe is old enough for that yet."

Caroline: "That's fascinating—and a little sad."

Zack: "I know. But the Universe doesn't cater to human feelings. And don't worry," he added with a grin, "we've got time—*perhaps more than 100 billion years before our Sun winds up as a black dwarf.*"

Caroline: "You said this applies to stars between 0.8 and 8 solar masses. What about larger ones?"

Zack: "Ah yes—the big ones. That's where things get wild."

Caroline: "Finally, we're getting to black holes!"

Zack: "Yes. *When very massive stars run out of fuel, their cores collapse—and the result is a supernova. For a few days or weeks, a supernova can outshine an entire galaxy.* In the core, protons and electrons are crushed together to form neutrons. What's left is a neutron star—so dense that a teaspoon of it would weigh as much as Mount Everest."

Caroline: "And black holes? Where do they fit in?"

Zack: "*If the collapsing core is more than about three solar masses*, even neutron degeneracy pressure—a quantum effect—can't stop the collapse. The core keeps shrinking until *it becomes a black hole: an object so dense that not even light can escape*."

Caroline: "That's unbelievable, and beyond intuition—I mean light can't escape? And what's this 'neutron degeneracy' thing?"

Zack: "Yes it is unintuitive, but remarkably general relativity predicts that gravity can bend light —and that's now been verified countless times."

Zack continues: "With respect to neutron degeneracy, that takes us into the other revolution in 20th century physics: quantum mechanics. You learned about the Pauli exclusion principle in chemistry, right?"

Caroline: "Yeah. Electrons, protons, and neutrons can't all occupy the same quantum state."

Zack: "Right. So in a neutron star, gravity tries to squash all the neutrons together, but quantum mechanics resists. That resistance is what we call *neutron degeneracy pressure*. But if the mass is too high, gravity wins—and a black hole forms."

Caroline: "And that's where *gravitational waves* come in?"

Zack: "Exactly. When massive objects like neutron stars or black holes accelerate—say, during collisions—they create *ripples in spacetime* called gravitational waves. These waves travel at the speed of light and stretch and squeeze space itself."

Caroline: "So they're like sound waves in the sense that they transport energy, but they don't need a medium?"

Zack: "Right. Gravitational waves move through the vacuum of space. And they were first detected in 2015 as faint ripples from a distant black hole merger. It was a huge confirmation of Einstein's theory.

Caroline: "Wow. That's a lot to take in. I definitely need to read more."

Zack: "I'll send you some links."

The call ends and Caroline turns to her father with a subtle smirk and a touch of sarcasm: "If it's OK with you, I won't ask about matter waves—at least not right away." Don lets out a very audible sigh of relief and picks up where they left off.

A 2-foot Long Piano String Produces a Middle C, Which Has a 4-Foot Wavelength

Her father, at last back in his element, explains that "When you press a piano key, let's say middle C, a small soft pad strikes a string, and the resulting vibrations—just like those produced with a tuning fork—press against the surrounding air molecules, creating tiny variations in pressure by increasing and decreasing their concentration. However, unlike the pure tone produced by a tuning fork, a piano string vibrates in a complex pattern with multiple frequencies."

Caroline responds, "So *when you press the middle C key, you're not just hearing a single tone?*" The inflection of her voice asked for confirmation of what she recalled from her previous conversations.

Don walks to the Fazioli, opening the lid and pointing to the soundboard and some of the cables associated with various notes. He plays a key and points out the obvious—because the ends of the vibrating string are stationary, longitudinal movement is prohibited, and consequently the wave doesn't travel.

He then sketches a figure that represents middle C vibrating at the first harmonic, i.e., in *its fundamental mode,* and confirms Caroline's understanding.

Don: "Although most of the sound you hear is in the fundamental, *a substantial fraction of the overall energy includes contributions from vibrations having frequencies that are integer multiples of the fundamental. For middle C the first overtone (second harmonic) is at approximately (261.6 x 2 =) 523 Hz, the second overtone at approximately (261.6 x 3 =) 785 Hz, and so on.* These *overtones*, or *higher harmonics*, enrich the sound, giving it depth and complexity. In addition—and this is important even though I won't elaborate—the sound would be very weak were it not for the fact that vibrations are transferred to a flat wooden soundboard that amplifies them and adds additional harmonics. That's perhaps not surprising: after all, there's a serious limit to how loud a sound you can achieve when a small felt hammer hits a metallic string."

Caroline: "What I think is fascinating is that the motion of a vibrating string is so much more complex than a wave that propagates in air."

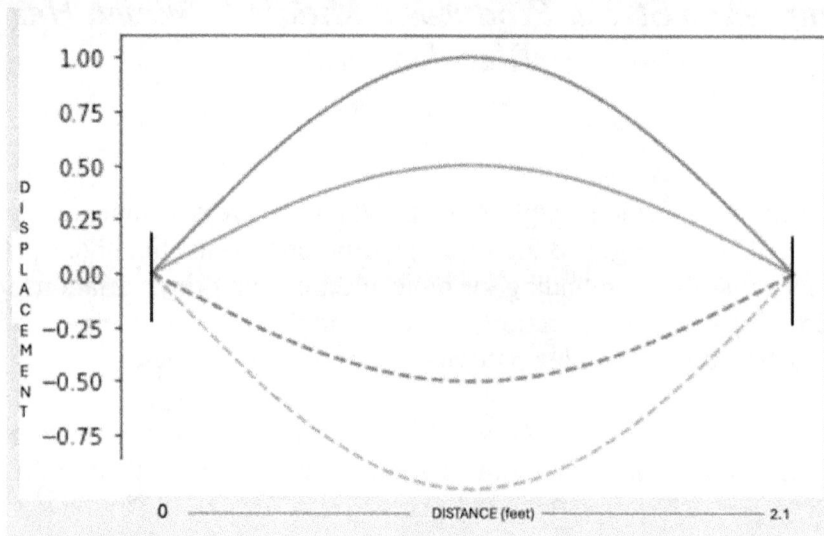

*Figure 8. A piano string vibrating in middle C's fundamental mode has fixed points, called nodes, that are approximately 2.1 ft apart, half the length of the **standing wave** on the string. Each curve represents a snapshot of **the fundamental** mode at different points in time. This fundamental vibration produces a **longitudinal sound wave** in the surrounding air, whose wavelength is **twice the length of the string.** When the string reaches its maximum displacement (the top curve in blue), it is momentarily at rest and exerts minimal influence on the surrounding air pressure. As it moves toward the equilibrium position, its speed increases, reaching a maximum there. This is where it contributes most effectively to sound wave generation by causing the greatest rate of air compression and rarefaction. The string then continues through the opposite phase of its motion, completing a half cycle when the node is at -1. This corresponds to a complete cycle of the resulting **longitudinal** pressure wave in the air.*

Don points out that the fixed ends of the string, *the nodes*, are approximately 2.1 feet apart when the string is straight but relaxed. When it vibrates, maximum displacement is reached by the antinode, which is at the center of the string.

Caroline, who recalls the relation between speed, frequency, and wavelength, and knows the speed of sound and frequency of middle C, points out that it has a wavelength of 4.2 feet. "So how can a 2-foot-long string produce a note with a wavelength twice as long?"

Don: "Great question. Let's look closely at the motion."

Pointing to figure 8, he starts by explaining, "When the antinode is at its maximum displacement, it is changing direction and is therefore momentarily motionless, exerting no pressure on the surrounding air. As the string moves from its maximum toward its relaxed position, it accelerates, starts to exert pressure on the surrounding air, and reaches maximum velocity as it passes through its relaxed, unstretched position. At that point, the string is straight and exerts maximum pressure on the surrounding air—and therefore air compression is also at a maximum."

Caroline: "So as the antinode moves from a displacement of 1 in the figure to a displacement of 0, air has changed from minimum compression to maximum compression."

Don: "Exactly. Can you tell me what happens next?"

Caroline: "As the antinode moves from 0 to -1, the surrounding air moves from maximum compression to minimum compression."

Don: "And..." (asking Caroline to complete the thought)

Caroline initially has a quizzical look, but it soon turns to one of satisfaction as she realizes that "as the string completes a half cycle, the corresponding pressure wave completes one full cycle."

Don: "Another way of looking at this is to realize that the sound wave is produced by the changing velocity of the string. So as the string velocity goes through one full cycle—from minimum (displacement +1) to maximum (displacement 0) back to minimum (displacement -1)—so does the corresponding pressure wave."

Caroline, with relief—and giddily tossing out some Greek: "Finally, now I understand how a string only λ feet long can produce a pressure wave 2λ feet long."

Don: "You should keep in mind, however, that this description is only an approximation for the vibration of a piano string..."

Caroline (interrupting): "Because, as you just said, real *piano strings vibrate with multiple overtones, meaning the motion isn't a perfect sine wave.*"

Figure 9. A vibrating string fixed at each end takes the form of a standing wave, in which each point moves only vertically. As the upper portion of the solid line moves down, and the lower portion moves up, the string assumes the shape of the dashed line. If the distance between the two end points is set at approximately 2.1 ft, the string, when vibrating as shown, will produce an air pressure variation with a frequency of approximately 525 Hz, which is close to the second harmonic of middle C (523.2 Hz) produced when the string vibrates in its second harmonic mode.

Some Waves Stand Still

Don emphasizes that each point along the string moves only in the vertical direction, and in that sense each point along the string is stationary. "This kind of *wave, in which the energy is confined to a physical object and doesn't move through space, is called a standing wave: it simply vibrates in place without traveling.*"

He then points to another diagram (Figure 9) that shows the second harmonic of middle C, but before he can continue, Caroline jumps in enthusiastically:

"Now I also understand why the fundamental frequency is the lowest possible frequency for a given fixed string length. This is the simplest possible standing wave for a fixed length: two nodes and one antinode in the middle."

Don, asking his daughter to extend her observation: "And what can you say about the second diagram?"

Caroline: "It has an additional node in the middle and an additional antinode: three nodes and two antinodes?"

Don: "What about the nth harmonic, where *n* is an integer?"

Caroline: "That one's easy: *n + 1* nodes and *n* antinodes."

Don continues with his explanation of the figure: "With the string vibrating at the frequency of its second harmonic, the upper portion of the solid line moves down, as the lower portion moves up, and the string assumes the shape of the dashed line. If the distance between the two fixed points is set at approximately 2.1 feet, the string, when vibrating as shown, will produce an air pressure variation with a frequency of approximately 525 Hz, which is the second harmonic of middle C."

Caroline: "It's obvious that they can't travel because they're fixed at both ends, but there's something about them that makes me feel uneasy—I can't quite picture how they form."

Don then, with some trepidation because he suspects that Caroline might not grasp the explanation he's about to give, writes an equation for the amplitude of two waves that not only vary with time, but are also moving through space in opposite directions along the *x*-axis. When they interfere with each other, amplitudes are subtracted so that the resulting amplitude is:

$$A(x, t) = \alpha\sin[2\pi f(t + x/v)] - \alpha\sin[2\pi f(t - x/v)]$$

where $A(x, t)$ is the wave amplitude at location x and time t, α is its maximum amplitude, f is its frequency, and v is its velocity (the speed of sound under standard atmospheric conditions).

He explains that at any given time the slope will be zero at the position at which the amplitude is maximum and adds that simple calculus can be used to find the slope. He knows that Caroline has yet to take calculus, so he calls it quits for the day, asking her to do some research and come back when she has found an expression for the position and amplitude of the antinodes.

A few weeks later she corners her father after dinner, notebook in hand, telling him that she took some Khan Academy lessons and had what she thought was the answer.

"I first used trigonometric identities to simplify the expression you gave me and found that:

$$A(x, t) = 2\alpha\sin(2\pi fx/v)\cos(2\pi ft)"$$

Then she says with some hesitancy and an uncharacteristic lack of confidence—one that often accompanies encounters with new and abstract concepts, "you'll have to check me on this, but when I differentiated the equation with respect to x (position) to obtain the slope, I found that:

$$S = (4\pi\alpha f/v)\cos(2\pi fx/v)\cos(2\pi ft).$$

"In order to obtain this result I had to use the fact that the derivative of a sine wave at a particular point is the value of a cosine wave at that point. I also started to wonder where that relationship, and a similar relation for a cosine wave, came from. It took me a while but I finally proved both relations."

Don: "I'm impressed. If anyone asked me to prove that the derivative of sine is cosine I doubt that I could do it: I never even asked the question. In fact, I think most people accept it—just as I think most people accept what they hear and what they see about most things—and never delve deeply, a tendency that might in fact have broad implications."

Caroline: "For example?"

Don: "For example, our understanding of complex issues like foreign policy can be distorted if we simply accept information at face value without critical examination and independent verification."

Then Don quickly returns to the topic by confirming that her answer was correct and asking if she could write an equation for the position x_a at which the amplitude is maximum. She stared at the equation a bit before recalling that "the slope has to be zero at the maximum, and that will happen when the cosine is zero."

Don: "And when will the cosine be zero?"

Caroline: "When its argument is $\pi/2$; i.e., when

$$2\pi fx_a/v = \pi/2$$

which means," she says with a feeling of satisfaction, "that since $v/f = \lambda$, $x_a = \lambda/4$"

Don then coaxes his daughter to develop the consequences further: "Can you now tell me the maximum value of the amplitude?"

Caroline has to think a bit but soon realizes that if she substitutes the expression for x_a into the expression for the amplitude $A(x, t)$, that should provide the answer:

$A(x_a, t) = 2\alpha\sin(2\pi f x_a/v)\cos(2\pi ft) = 2\alpha\sin(\pi/2)\cos(2\pi ft)$ or

$A(x_a, t) = 2\alpha\cos(2\pi ft)$

which says that the amplitude at the antinode varies sinusoidally with time.

Don then asks her for a coherent summary of her results.

Caroline, sounding more confident and trying to summon up all the new words she learned:

"The amplitude of the standing wave at the antinode decreases sinusoidally with time, passes through its equilibrium ($A(x_a, t) = 0$) at $t = 1/4f$, and continues on to its other maximum (the lowest curve in Figure 7), which occurs at $t = 1/2f$."

Don: "Wow. My next question was going to be about time—how did you determine that the time it took the antinode to move from its maximum to the equilibrium position was one divided by four times the frequency?"

Caroline: "The antinode passes through its equilibrium position twice, once on its way to the second maximum (Figure 8) and then again on its way back to the initial maximum. Consequently, the first time it passes through equilibrium it has completed a quarter of a cycle, which means that $2\pi ft = \pi/4$, giving $t = 1/4f$."

Don: "This is in contrast to the pressure wave, which is at its minimum when the antinode is at its maximum displacement."

Caroline: "Yes, I get it—when the antinode passes through its equilibrium position, the string is stretched straight, moving with maximum velocity and exerting maximum pressure on the air molecules. Whereas when the string is maximally displaced, it's changing direction, so it's instantaneously motionless and therefore exerting essentially no pressure on the air. This is essentially what we said earlier."

Don: "Positions between the node, where the ends are fixed, and the antinode, where displacement is maximum, follow the same type of motion, except the amplitudes aren't as large. For example, at $x = \lambda/8$:

$A(\lambda/8, t) = 2\alpha\sin(\pi/4)\cos(2\pi ft) = \alpha\sqrt{2}\cos(2\pi ft)$.

"This shows that each position along the string follows sinusoidal motion with the same frequency as the antinode, but with a smaller and diminishing amplitude as the position gets closer to the node.

"It's the same type of behavior exhibited by a tuning fork, which also supports standing waves, but for a tuning fork the node is the joint where the prongs meet, and the antinode is at the free end of the prongs, where the vibration is at its maximum."

Don, as usual, is exhausted by his daughter and calls it quits for the day but encourages her to think about generalizing the expressions for the locations of nodes and antinodes for higher-order harmonics.

It is now two Saturdays later and Caroline is eager to speak with her father, but there's no answer when she knocks on the door of his study.

"Mom, have you seen Dad? I wanted to show him some results I found for a vibrating string."

Vibrating string catches Mary's attention: "I had an entire course on that—it even included vibrating surfaces," she says before realizing she hasn't responded to Caroline's question. "He's meeting with some members of the orchestra and won't be back until later this afternoon."

Caroline, eager to show her results, opens her notebook and shows it to her mother.

"I used calculus to analyze a standing wave, but there's a much easier way. I had derived the equation for the amplitude, or displacement:

$$A(x, t) = 2\alpha \sin(2\pi x/\lambda)\cos(2\pi ft)$$

and what I want to find are expressions for the positions of the nodes and antinodes for a string whose two ends are fixed a distance *L* apart when the string is relaxed and fully extended. The key is that for the *n*th harmonic the wavelength must satisfy *λ = 2L/n*.

Mary: "Why is that?"

Caroline: "Because a standing wave on a string occurs when an integer number of **half wavelengths** fit **exactly** into the string of length. Since each loop of the standing wave (one peak and one trough) represents half a wavelength ($\lambda/2$), if there are n loops (or n half wavelengths) *L = nλ/2*.

"Then when we substitute the expression for λ into the expression for the amplitude, we find that the spatial dependence of the standing wave is given by:

$$A(x, t) = 2\alpha \sin(\pi nx/L)\cos(2\pi ft),$$

and since the amplitude is zero at the nodes

$$\pi nx/L = m\pi \qquad m = 0, 1, 2, \dots, n$$

$$x = mL/n \qquad m = 0, 1, 2, \dots, n$$

"So the nodes for the nth harmonic occur at:

$$x = 0, L/n, 2L/n, \dots, L$$

"On the other hand the antinodes, the points of maximum displacement, occur where the sin is ± 1. That means

$$n\pi x/L = (m+1/2)\pi, \qquad m = 0, 1, 2, \dots, n-1$$

or

$$x = (2m+1)L/2n, \qquad m = 0, 1, 2, \dots, n-1$$

Consequently, the antinodes for the nth harmonic occur at:

$$x = L/2n, 3L/2n, 5L/2n, \dots, (2n-1)L/2n"$$

Mary: "Wow, that's really impressive. Did you do that on your own?"

Caroline: "Pretty much. I showed Noreen what I had done with Dad. She was captivated, and we both started thinking about harmonics. We were both stuck, but you know Noreen—it wasn't long before she realized that wavelength and the length of the string must satisfy $\lambda = 2L/n$. After that we quickly and independently found all the relationships."

Mary was struck by her daughter's derivations; after all, she was still only in high school.

Mary: "We're now ready to discuss why instruments playing the same note sound so different, and related questions. Let's go to the piano.

"But before going further, I should comment on terminology so I don't again make the mistake of using unfamiliar words."

Mary reminds her daughter that *starting at any particular note, say middle C, and counting keys to the right, there are 7 white keys and 5 black keys before the same note—at double the frequency—is reached.*

"Because the same note is reached on the 8th white key, the set of 12 keys is called an octave, from the Latin meaning eight—although," she adds with a mild grin, "it could just as well have been called a *duodecim*."

Caroline, remembering her aunt's love of language, quips, "That's a word worthy of your sister."

Mary: "I guess it runs in our family."

Mary continues: "Furthermore, instead of saying things like the C just above middle C or two Cs above middle C, or the A above middle C, there's a simple notation that avoids all those words. Middle C is referred to as C4. The three Cs below it, starting at the lowest pitch, are called C1, C2, and C3, and the four above it are C5, C6, C7, and C8. And the A above it is...?"

Caroline: "I guess that would be A4. But why is C4 called middle C? It's not in the middle. In fact, there is no middle because there are 8 Cs on an 88-key piano."

Mary: "Let's not go too far afield. I'll just say it has to do with staves: treble clef, bass clef, and grand staff. I'll let you research the answer for yourself. Just know the notation and recognize that it's useful because, for example, saying 'C5' is more efficient than saying 'the C one octave above middle C.'

"Now, where was I? OK, *if a middle C is played on an instrument, or is sung, then the resulting sound will consist mostly of the 262 Hz wave—that is, most of the energy of the perturbed air molecules will be in that energy wave. But, as I mentioned before, some of the energy will be in the higher frequencies: some in C5, which has a frequency of 2 × 262 Hz, some in C6 at 3 × 262 Hz, and so on.*

"Now you can understand why when you play middle C and the C just above it simultaneously, the sound is so pleasant, so harmonious. It's because when middle C is played, the 262 Hz vibration is reinforced by all its overtones."

Caroline: "Are there other waves vibrating at 4 × 262 Hz, and 5 × 262 Hz? When does it stop?"

Mary: "The answer to your first question is yes. The empirical answer to your second question is that it stops when the amount of energy is so small that you can no longer detect it—that's at approximately twenty kilohertz (20 kHz) for normal human hearing."

Caroline: "My teacher said that bats—as well as many other animals—hear much higher frequencies than humans. In fact, she said that bats see using sound. I didn't understand that, but I didn't ask."

The Same Note Played on Different Instruments Sounds Different Because Different Instruments Produce Different Harmonic Spectra

Mary: "Yes, that's true—*bats emit high-frequency sounds and are able to detect shifts in frequency, intensity, bandwidth, time delay, and so on, which allows them to determine properties of objects in their environment such as size, shape, and distance with remarkable precision.* The physics is very interesting, and I'm sure the neurobiology is as well. Remember, you have an uncle at Los Alamos who's a physicist, and an aunt who's a biology professor, and I believe your aunt taught a course on animal senses—you might want to pick their brains if you're curious. But I suspect our conversation about sound should keep you occupied for a while."

Caroline responds with an "I'm not so sure about that" smugness, which is sometimes displayed by talented teens.

Mary continues, unfazed, to arrive at her main message: *the fraction of the total energy in each harmonic—that is, the way energy is distributed among the various overtones—is generally different for different sound sources. That's why different instruments playing the same note sound different.*

Mary: "Don, would you like to chime in?"

Don, to his wife: "Sounds like you caught the pun bug, but yes, I'll add my voice to the conversation"

"*For middle C on a piano, some 60–70% of the energy will be carried by the fundamental frequency—that is, by the lowest frequency that can be generated by the middle C string. Approximately 20–30% will be carried by the first harmonic, and perhaps 5–10% by the second harmonic. For the trombone, the energy would be shifted a bit toward the higher harmonics something like 50–60%, 25–35%, and 10–20% in the fundamental and the first two harmonics.*"

Don emphasizes that the exact numbers depend on many factors, including the quality of the instrument, the material used to make it, the performer, and the energy of the sound.

Caroline: "What about the difference between the male and female voice?

Don: "As you might expect, the female voice has a relatively lower percentage of energy in the first harmonic and, more generally, is about an octave higher than the male voice.

"The human vocal system, which evolved by natural selection, is much more complex than any instrument constructed by mere mortals. It's affected by many parts of your anatomy—so a good answer to your question requires a separate and extended conversation."

Caroline: "You're not kidding, especially if you try to explain how natural selection influenced the development of vocal systems—I'd love to hear about that."

Don: "I'm sure you would, but you'll have to ask someone else—your Aunt Zoe or a teacher. It's difficult enough for me to describe the human vocal system as it is, let alone how it got that way. Even as a professional musician, I don't know enough to give more than a general answer to your question. A professional singer can tell you more, and if you're interested, I can introduce you to a colleague who is an accomplished soprano."

Caroline: "OK, maybe when I have more time."

Don: "I couldn't agree more, so let's take the plunge.

"When you sing vowels like 'aaah' or 'eeee,' your vocal tract acts like a custom instrument. The shape of your mouth and the placement of your tongue change which frequencies are amplified, just like changing the length of a trumpet affects its pitch."

Caroline: "So my mouth and throat are like a musical instrument?"

Mary: "Exactly! Think of them as adjustable resonators. *A wide mouth emphasizes low frequencies, giving 'aaah' its rich tone, while a narrow shape amplifies high frequencies, creating the bright sound of 'eeee.'* It's sort of like how different instruments can play the same note, but they sound different because of their unique resonance characteristics. The violin, trumpet, and piano all have different resonance behaviors that color the same pitch differently."

Delighted by this scientific insight into her musical world, Caroline asks eagerly, "Can you tell me more about overtones and resonance?" Then she realizes that she really isn't sure she understands the meaning of *resonance*—even though she just used the word.

Caroline: "I'm not really sure I know what resonance means," she admitted, eager to fully grasp the concept her mother was explaining.

Mary: "I'm glad you asked that question—resonance is a subtle concept, and a lot of people use words without having a clear understanding of their meaning, especially when the words refer to a concept." Then, with a jocular tone and a playful expression on her face, she added, *"Understanding what you don't understand is a great first step toward understanding."*

"Resonance refers to the reinforcement or amplification of certain frequencies when a system is set into vibration. A common example is pumping a playground swing—you time your pushes to match the natural swing frequency, and as a result the arcs of the swing grow larger and larger. With musical instruments and our vocal tracts, different structures resonate at different frequencies based on their size and shape. For example, think of blowing across the top of a bottle—you can get different pitched tones depending on how much air is in the bottle, right?"

Caroline nodded, recalling that experience.

Her mother continued: "The air column trapped in the bottle resonates strongest at a particular frequency determined by the bottle's volume. The same principle applies to instruments—and the same applies to the human body: they act as resonators that selectively amplify different frequencies.

"When you change the shape of your mouth and throat in order to sing different vowels, you're changing the resonant frequencies being amplified, which alters the tone, color, or timbre of your voice—even when the pitch stays the same.

"In addition, *superimposed on the fundamental frequency and its overtones is a much higher and steady frequency in the range of 4–8 Hz. It's called vibrato and first became prominent in Italian opera.* And remarkably, although it involves controlling airflow, relaxation of the vocal cords, and resonance produced by the chest and mouth, it occurs naturally in the voices of accomplished singers. In fact, effort is required to suppress it, and although it generally enriches sound, some portions of some songs are best produced without it."

Mary could see understanding dawning on her daughter's face as she began to grasp, even if ever so slightly, the complex interplay of resonances that give voices and instruments their unique sound, and the fascinating connection between physics and physiology that brings exhilaration to millions of people every day of the year, every year without stop.

Now, somewhat fatigued though feeling gratified, Mary addresses her daughter head-on: "Caroline, it's been fun, but now it's time for your homework, and your father and I would like to simply relax—we've had a tiring day and we're ready for a martini and anything television has to offer."
Caroline: "Mom, what's a martini?"

Figure 10. Dall-E rendition of Caroline musing over why the sounds of music are so moving. The left panel displays waves of energy driven by cycles of compressed and rarefied air; the right panel shows that Caroline perceives these waves as music. She's struck by how the brain turns something apparently so simple as air pressure variations, into such a deep emotional experience, and wonders whether physics and mathematics will ever provide a complete answer.

Conversation 3

Planes, Tops, Gyros and the Apparent Defiance of Gravity

During Caroline's trip to the airport to visit her uncle, her mother struggled to answer her question about why planes don't fall. As they discussed the physical principles governing flight, Mary mentioned the Bernoulli principle, opening a brief discussion about one of the most remarkable families in the history of science. The conversation reminded her of how mathematical discoveries made centuries ago—without any practical application in mind— eventually provided the foundation for technologies that wouldn't emerge for another two centuries. As the discussion broadened to the topic of flight, Caroline received her first lesson in biological engineering: the recognition that nature is the greatest engineer of all.

Newton's Third Law and Pressure Differentials Help Keep Kiloton Planes in the Air

It's a bright, sunny Sunday morning in New York, and Caroline and her parents are off to an early start—out of the house by 8 a.m. and on their way to LaGuardia Airport. This will be Caroline's first trip to Santa Fe, New Mexico, where her Uncle Zack lives, and she'll be visiting without her parents. Although Caroline will turn 15 in just a few weeks and has flown before, she has never flown alone. Strangely, she has never thought much about how planes fly. But this time, perhaps because she's anxious, it will be different.

"Hey Mom, why don't planes fall?"

Caroline's mother, still half asleep and a little uneasy about her daughter flying alone, isn't ready for deep thinking. She quips absentmindedly, "Why don't birds fall?"

"Because they flap their wings," Caroline replies without hesitation. "And besides, birds are so much lighter than planes. I read that 747s can weigh 400,000 pounds!"

Mary, now fully awake, nods as she considers the marvel of modern aviation.

"You're right. A 747 weighs hundreds of thousands of pounds, has millions of parts working together, flies six miles above the Earth at nearly the speed of sound for hours, and can travel more than 10 million miles in its lifetime without a single fatality. Compared to that, what birds do seems simple. But in reality, planes and birds rely on the same basic physics—just on a vastly different scale. The real miracle is that humans figured out how to work together to make it possible."

Caroline catches something that surprises her. "Nearly the speed of sound? I didn't realize planes flew that fast."

She quickly does the math in her head: $760 \times 0.8 = 700 \times 0.8 + 60 \times 0.8 = 560 + 48 = 608$ miles per hour.

Mary is impressed. "You even know the speed of sound! But keep in mind, *760 mph is the speed of sound at sea level—it actually decreases at higher altitudes* because...?", Mary prompts her daughter for an answer.

Caroline says, "The air is thinner?"

Mary replies, "Good guess; but that's a common misconception. *The speed of sound depends mostly on temperature,* and the temperature at 35,000 - 40,000 feet, which is where most commercial airlines fly, is considerably less than it is at sea level."

Caroline asks, "Shouldn't the temperature increase if you're getting closer to the Sun?"

Mary answers, "Not at airline altitudes, and ordinarily I'd ask you to research the answer for yourself, but since we're in a car, I'll give you the broad brush explanation."

Mary continues, "Near the ground, air is warmed by heat radiating from the Earth's surface, which absorbs energy from the Sun."

Caroline says, "So the air isn't directly heated by the Sun—its temperature is lower because of its distance from the heat source?"

Mary replies, "Part of the temperature difference comes from the drop in pressure, which causes the air to expand and cool.."

Caroline asks, "Why does the pressure drop?"

What's going through Mary's mind right now is that some people might have been satisfied with the pressure drop answer, but she's sure Caroline will probe at least two more levels, and that will lead them into thermodynamics, which she won't be able to explain, in part because the subject is subtle and in part because Caroline hasn't had enough physics—and because they're in a car with no blackboard.

Mary replies, "The pressure drops because there's less air above you at 40,000 feet than at sea level."

Caroline: "OK I knew that, but why does the reduced pressure mean a reduced temperature?"

Mary explains, "Because when the *pressure is lower, the air expands and as it expands, it cools—* at least under certain conditions." She then adds with a minor feeling of frustration because she always loved thermodynamics and was uneasy about oversimplifying: "If you'd like to delve more deeply than that, feel free to ask your uncle— I'm sure he'd be happy to provide some guidance."

Mary continues, "I'll add one more piece of information, just to keep you busy. You asked about distance from the Sun. What I've described are processes that occur in the *troposphere*, which *is the atmospheric layer in which commercial airlines fly. In the next layer up, in the stratosphere, temperature actually starts to increase again because the ozone layer absorbs ultraviolet radiation from the Sun."*

Caroline concludes, "I guess the bottom line is that the speed of sound in the troposphere decreases because the rate of energy transfer decreases, and that decrease results from a reduction in the speed with which air molecules move, and reduction in speed results from a reduction in temperature which is a consequence of the pressure differential between the ground and the plane."

Mary, half tongue in cheek, says, "Simple."

Though satisfied with the explanation, Caroline circles back to her original question. "OK, but *what exactly keeps a 400,000-pound plane in the air?"*

Mary smiles. "It's a combination of wing shape, the angle of attack—the angle the wings make with the airflow—and differences in air pressure above and below the wings."

"That's a lot to take in, especially this early in the morning!" Caroline laughs.

"Let's start with angle of attack," Mary suggests.

She guides Caroline through a simple experiment. "Open the car window, keep your fingers together, and place your hand outside—but keep it close to the car. Hold your palm flat, parallel to the ground, and then slowly tilt it upward."

Caroline immediately feels the change in pressure. "Whoa! When my palm is perpendicular to the ground, the air pushes it backward really hard. But as I tilt it forward, the backward push decreases, and I start to feel an upward force too. When it's completely flat, there's barely any push at all."

Mary nods. "Exactly. *The upward force is called 'lift,' and the backward force is 'drag.' Planes rely on both, but they want to maximize lift while minimizing drag."*

"So *changing the angle of attack changes the ratio of lift to drag?"* Caroline asks.

"Yes," Mary confirms. "And what happens if the car accelerates while you keep the same angle?"

Caroline: "The force on my hand would increase. So both lift and drag depend on the angle of attack and velocity."

"Exactly", Mary confirms, "but *there's something else that's interesting about lift: it depends on the velocity squared*."

Caroline: "The square of velocity? Why?"

Mary grins. "That explanation is a bit more technical. Should we add it to our 'rainy Saturday' list, or do you want to ask your Uncle Zack in Los Alamos?"

Caroline smirks. "I'd rather know now so I can think about it on the plane!"

Mary relents. "Alright. You know kinetic energy is associated with motion, right?"

Caroline: "Yes."

Mary continues, "Well, when a plane moves forward, it collides with air molecules that were previously stationary. These molecules transfer kinetic energy proportional to mv^2 where v is the plane's speed and m is the mass of a molecule."

Caroline, catching on, interjects. "Wait—when my hand was perpendicular to the air, I felt the strongest push. Doesn't that mean the energy transfer also depends on the angle at which the molecules hit the wing?"

Mary: "Yes, exactly. The force is still proportional to v^2, but the actual lift depends on the angle of attack, because not all of that energy is directed upwards. *As the plane moves, it deflects air downward, and by Newton's Third Law, the air pushes back, lifting the plane.*"

Caroline: "So the more air molecules hitting the wings per second, the more energy is transferred, and the more lift is generated?"

Mary confirms Caroline's understanding: "That's right. And because there are fewer molecules in thinner air, lift also depends on air density. If you stick your hand out of a car window in thin air, the force on your hand would be much weaker."

Caroline, seeking to confirm that her understanding is correct: "So *lift is related to air density, speed squared, and angle of attack*?"

"Yes," Mary confirms. "But there's a catch. If the angle of attack gets too high—say, 15 to 20 degrees—the airflow becomes turbulent, and lift decreases instead of increasing. That's called a stall."

Caroline, again confirming: "Oh, so planes don't just use the maximum angle of attack—they use the optimum?"

Mary: "Exactly. Planes take off with a relatively high angle of attack to maximize lift, but once they reach cruising altitude, they reduce it and compensate by increasing speed."

Caroline: "And higher speeds create a bigger pressure difference between the top and bottom of the wing?"

"Yes," Mary replies. "But the relationship is complicated. The shape of the wing also plays a major role, and that involves some pretty advanced math—like Bernoulli's principle and something called the Kutta-Joukowski theorem."

Mary trails off, relieved that Caroline doesn't ask for more details. There will be plenty of time in the future for that level of depth—once she's learned the math to match her curiosity.

Caroline leans back, thoughtful. "OK, that actually makes sense. I'll think about it on the plane… and maybe I'll ask Uncle Zack about the math."

Mary smiles. "That sounds like a great plan."

Daniel Bernoulli: A Legacy of Scientific Genius

But just as Mary was about to relax, Caroline remarked that her physics teacher had described something called the Bernoulli principle, which she didn't fully understand, but which seemed somehow relevant.

Caroline: "Dr. Khan said that, in addition to Newton's Third Law, the generation of lift involves another principle—named after someone, of course—this time a guy named Bernoulli. It explains how differences in airspeed create pressure differences,"

Mary nodded. "*The Bernoulli principle helps explain part of how lift is generated.* As air flows over the wing, its shape—the *airfoil*—and its angle of attack cause the air to move faster over the top surface and slower beneath. According to the Bernoulli principle, faster-moving air has lower pressure, while slower-moving air has higher pressure. This pressure difference creates an upward force, which we call lift. But that's only part of the story. Lift is also generated because the wing deflects air downward, which creates an equal and opposite upward reaction force in accordance with Newton's Third Law. Together, these effects keep an airplane in the air."

She continued, "Both principles work together. *At lower angles of attack, the Bernoulli effect dominates because the airflow remains smooth and streamlined over the wing. At higher angles of attack, the role of Newtonian mechanics becomes more significant as the wing directly pushes against the air.* It's the combination that makes lift efficient under different flight conditions."

Then Mary added, with a pointed tone that sounded almost insulted, "And, by the way, Daniel Bernoulli—or 'this guy,' as you call him—was part of one of the most remarkable families in the history of science. He had no fewer than eight close relatives who made major contributions to mathematics and physics."

She went on, "And beyond the scientific importance of the Bernoulli principle, there's something else to appreciate. When Daniel Bernoulli published his work on fluid dynamics, no one imagined it would lay the groundwork for technologies like the airplane and the carburetor—technologies that wouldn't exist for another two centuries. So, the next time you hear someone say mathematicians waste their time proving theorems of no practical value, remember that *one of the key enablers of modern transportation was discovered by someone simply following his intellectual curiosity, with no idea how important it would become long after he was gone.*"

Mary, clearly on a roll, couldn't help adding more. "The Bernoullis also made extremely important contributions to probability and statistics. Their work laid the foundation for sampling theory, which allows us to draw reliable conclusions about huge populations based on small samples. It's hard to overstate how essential that is in everything from science to social policy. And, just to put a punctuation mark on how brilliant Daniel Bernoulli was—his PhD wasn't even in mathematics; it was in anatomy. That was due to pressure from his father, who was also a famous mathematician, though apparently not of Daniel's caliber. Tragically, there was serious animosity between them. I don't know the full story, but it might make interesting reading."

Birds and Planes: Different Objectives, Same Physics

Caroline looked at her mother with a sly smile. "I guess you don't want to go into detail about his theorem?" she teased.

Mary replied, chuckling, "I realize my explanations are simplified, but I think—at least I hope—they capture the most essential parts of the physics."

Caroline said, "Yes, I understand much better than I did a little while ago why planes stay aloft. But I can't help thinking about something that's only marginally related but also very important: the airplane is an example of humans outperforming nature. We can fly much, much faster and much, much farther than birds."

Mary agreed, with a hint of sarcasm in her voice. "Yes, the airplane is an example of human engineering outperforming evolutionary engineering—all it took was a little planning and some money."

Caroline: "Mom, that's silly. It's just like you to mention money. What this conversation taught me is that it took extraordinary creativity, ingenuity, and the human capacity for organizing."

Mary: "I know," her mother replies. "I was trying to be funny, and I'm just starting to wake up."

Don: "The comparison between birds and planes is interesting, but I think it might be missing something. I'm not a biologist, but it seems to me that *the flight objectives for birds and planes are different.* Evolution has optimized bird flight for energy efficiency and maneuverability, whereas planes are optimized for speed and endurance. So, it's not a matter of better or worse—the fact is that they have different design objectives, and those objectives are met quite well by both."

Mary, now fully awake, agrees with her husband but Caroline, with typical enthusiasm, cuts in.

"I still don't understand the relation between the way birds fly and the way planes fly—planes don't flap their wings"

Mary: "Now who's being silly?"

Caroline: Well, what's the answer?"

Mary: "The answer, I believe, is that although planes don't flap their wings, the physical principles underlying bird flight are the same as those used by planes."

Turning to her daughter, she asks, "What was one of the key principles involved in lift, from our conversation a little while ago?"

Caroline: "I'm not sure."

Mary: "Actually, I used a word that was a little bit misleading. It's not a principle—it's a law."

Caroline: "Oh, you mean Newton's Third Law?"

Mary: "Yes, exactly! When a bird takes off, it flaps its wings, pushing the air back and down so that it can generate lift, the embodiment of Newton's Third Law, which says that for every action, there's an equal but opposite reaction. In addition, *a bird can also rotate its wings to vary the angle of attack and thereby modulate the force determined by Newton's Third Law.* Since a bird is 100,000 times lighter than a large plane, its speed doesn't have to be very great to generate the required amount of lift, as long as its wings are large enough—and evolution—that great engineer, as your aunt refers to nature—has made sure that they are."

Caroline: "Mom, what does 'evolution—that great engineer' mean?"

Don: "It means just what the example your mother gave illustrates. Natural selection has designed living organisms, and in some cases optimized their design, so that they can adapt to various environments. In fact, biology—and evolutionary principles in particular—have been so successful that an entire field is emerging called biologically inspired engineering."

Mary: "*Leonardo da Vinci might have been the first person to realize that natural processes can inspire and help improve engineering.* One example is his design of a flying machine based on detailed observations of bird anatomy and flight patterns; another is his submarine. *Biomimetics* is now a common approach to all kinds of engineering objectives, from the study of hummingbirds for the design of flying robots, to the development of artificial photosynthesis for converting light into chemical energy."

Caroline tells her parents that she's fascinated by what they just told her, even as she continues to ponder ways in which gravity is defied.

A Spinning Top Generates a Gravitational Torque that Changes the Direction of its Angular Momentum, Causing it to Precess

Caroline said, "I've now encountered two examples of what I like to call gravity defiance: the fluid in a straw and the airplane—and both are related to pressure differentials. I can't help feeling that there are many other ways in which gravity can be defied."

Mary replied, "There are, and I'm sure you've come across them."

Caroline pondered, but nothing came to mind.

Mary added, "How about the spinning top? As long as it spins, it doesn't fall."

Don interrupted to tell his wife and daughter, who had been totally engrossed in conversation, that they had just arrived at the parking lot and that additional questions would have to wait. "It's going to take a while to get through security, and it's still early, so let's find a restaurant and we can continue the conversation there."

Now at the restaurant, Caroline said, "I can't imagine why the top doesn't topple immediately."

She was delighted when Mary responded, "The mechanics of this phenomenon are beyond what you've learned in school, but it might be fun to try to understand why a spinning top doesn't fall."

Mary continued, "Let's start with something simple. In your first-year physics course, you probably learned that the *linear momentum of an object moving in a straight line is its mass multiplied by its velocity, and that the force exerted on an object equals the rate at which its linear momentum changes*."

Caroline replied, "Yes, I did. And I also learned that *when an object moves in a curved path around an axis, it possesses angular momentum, and that the direction of the angular momentum vector is determined by the right-hand rule and points along the axis of rotation*. In this case, force is replaced by *torque*, which is the force on the object multiplied by its perpendicular distance from the axis of rotation. *Torque changes the angular momentum of an object, either in magnitude, direction, or both*."

Mary: "Good, let's build on that. Did you learn about something called the center of mass?"

Caroline, "Very little— all I remember is that the translational motion of an extended object, the part that's not related to rotation of the object, can be described by assuming that its entire mass is concentrated at a single point called its center of mass. But honestly, I didn't quite understand it."

Mary: "That's OK, let's try a simple example. Imagine a straight object made up of N tiny masses, $m_1, m_2, m_3, ..., m_N$. The masses can, for example, be atoms or molecules, and they're positioned at $x_1, x_2, x_3, ..., x_N$, along the length of the object, where N is very large. Then the center of mass is defined as

$$r_{cm} = \frac{\sum_{i=1}^{N} m_i x_i}{\sum_{i=1}^{N} m_i}$$

where the sum in the denominator is the total mass of the object.

"In words, *the center of mass is calculated by taking the weighted average of each mass's position: total (mass × position) divided by total mass*."

For the simplest situation,

$m_i = m$ for every mass, and the expression for the center of mass becomes

$$r_{cm} = \frac{m \sum_{i=1}^{N} x_i}{Nm} = \frac{\sum_{i=1}^{N} x_i}{N}$$

"In other words, *for a uniform object, the center of mass is at the average position of the masses.* For example, a uniform symmetric object like a straight rod will have its center of mass at its midpoint This is the simplest situation. If you go on to study physics in college you'll learn how to do similar calculations for solid 3-D objects with continuous mass distributions."

Caroline said, "The result makes sense. What about something more complex, like a spinning top?"

Mary replied, "Yes, that's the question. But before we delve into it, I need to discuss the different symbols that are used to describe a vector."

Caroline's expression was a mixture of frustration and anxiety, reflecting a suspicion that she and her mother were about to get involved in a morass reminiscent of their earlier discussion about units.

Mary (noticing that her daughter seemed uneasy): "You look a bit anxious, but there's no need for concern. The ideas are straightforward."

Mary continued, "If I were writing at a blackboard, I would place an arrow over a symbol to represent a vector. However, if I were at a word processor, typing a symbol with an arrow over it would require several keystrokes, so a different representation is easier. The method of choice is to use a symbol written in boldface.

"Now let's consider a spinning top having a mass m and a central axis of length ξ, from its tip to its center of mass. The central axis is referred to as an axis of symmetry because rotation about it leaves the appearance of the top unchanged. Initially the top is upright, and its symmetry axis is perpendicular to a rough horizontal surface, so its tip is at height 2ξ above the surface.

Mary stops to sketch a detailed vector diagram.

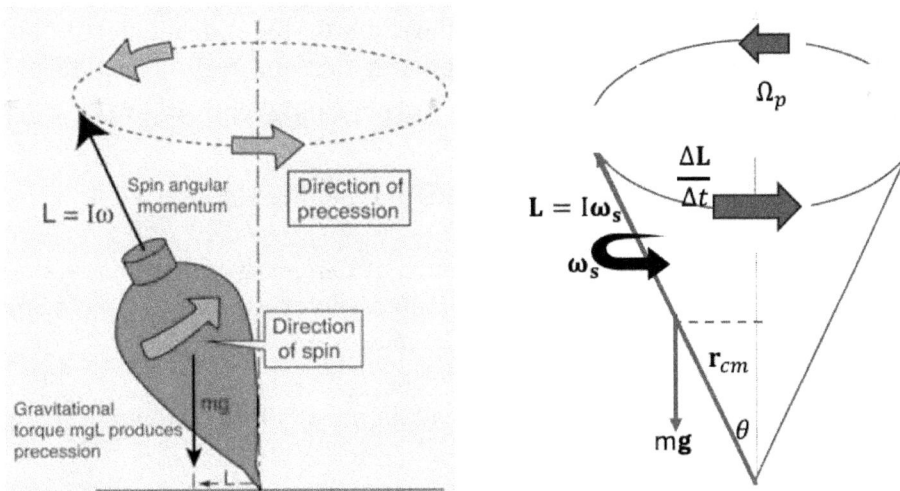

Figure 1 Left: A general overview of a precessing top. Right: A detailed vector diagram illustrating the forces on the precessing axis of symmetry. A gravitational torque of magnitude $\tau = mg\, r_{cm} \sin\theta$ acts in a counterclockwise direction parallel to the floor to change the direction but not the magnitude of the spin angular momentum. The result is precession at a rate $\Omega_p = \frac{\tau}{L}$ of the top's axis of symmetry, as shown from a top down perspective.

Caroline: "Why are we talking about a rough surface? In elementary physics my teacher usually talks about frictionless situations and vacuums."

Mary: "Good question. For some situations, such as in Galileo's experiments, for example, frictionless surfaces and vacuums are useful—they simplify the math and isolate principles. *If Galileo hadn't thought about objects falling in a vacuum, he probably wouldn't have anticipated what came to be known as Newton's first two laws of motion.*

"*For a spinning top, however, friction is essential: without it, the bottom of the top would simply slip and the top wouldn't remain upright.* So friction provides the stability needed for the top to precess instead of just toppling over."

Caroline: "Precess?"

Mary: "Sorry—*precession is the circular motion of the top's axis of symmetry around the vertical axis.*"

Caroline: "Yes, I know that, but why doesn't it fall, and what determines its precessional velocity and the angle of its tilt?"

Mary: "Let me try to explain. First, let's define the important quantities and their symbols. The top's center of mass, r_m, is at a distance ξ from each end, and its initial angular momentum is $L = I_s$

69

ω_s where I_s is the moment of inertia about the spin axis. But there's also another moment of inertia perpendicular to the spin axis, which is different than the spin moment of inertia. Calculating a moment of inertia is a bit complicated, but I can give you the result for the spin angular momentum of a solid cylinder—which is sometimes used as an approximation for a top, depending on its shape. If R is the radius of the cylinder (perpendicular to its spin axis) then $I_s = \frac{1}{2}mR^2$."

Caroline: "I'm a bit lost. Could you give a simple explanation of moment of inertia?"

Mary: "Yes, I should have defined it. Put simply, you can think of *moment of inertia as a measure of how difficult it is to change the rotational motion of an object that is rotating around a specific axis. It's essentially the rotational equivalent of mass in linear motion.*

"Here's another way to think about it. When you hold a small dumbbell close to your chest, turning around repeatedly isn't difficult."

Caroline: "It isn't? I would think you'd get pretty dizzy."

Mary: "OK, smart aleck, let's set dizziness aside and focus on energy expended. It's considerably less difficult than turning around when holding the dumbbell in outstretched arms. And the reason is that with outstretched arms, the mass is farther from your axis of rotation, increasing the moment of inertia."

Caroline: "So the heavier the object and the further it is from the axis of rotation, the more difficult it is for the object to rotate—it has, in effect, more rotational inertia."

Mary: "Yes, I couldn't have said it better. But how you determine the exact dependence on those quantities is another whole story, which you're not quite ready to learn about—at best, it would be a slog."

Caroline: "OK, I'll accept the expression you gave—at least for now."

Mary: "Now let's consider the angular velocity of precession when the spin axis is tilted at a fixed angle θ with respect to the vertical about which it is precessing."

Caroline, interrupting: "I've seen pictures of the Earth spinning about a tilted axis, but the axis is bobbing up and down."

Mary: "Yes, that slight up and down motion of the tip of the axis is called *nutation*. I'm neglecting that, in other words I'm assuming that θ doesn't change. The angular velocity about the vertical—let's call it $\boldsymbol{\Omega}$, will have two components: one along the spin axis ($\Omega \cos\theta$) and one perpendicular to the spin axis ($\Omega \sin\theta$). So what's the total angular velocity?"

Caroline looks puzzled.

Mary: "Think. Let's start with the component along the spin axis."

Caroline:" OK, I think I get it. The angular velocity along the spin axis should be $\omega_s + \Omega \cos \theta$ and the velocity perpendicular to the spin axis should be $\Omega \sin \theta$."

Mary (getting tired and now eager to end the back and forth): "And therefore the total angular momentum is $I_s(\omega_s + \Omega \cos \theta) + I_p \Omega \sin \theta$. The second term is the angular momentum of precession.

"When the top is set in motion, its central axis will be, or will quickly become, misaligned from the vertical—however slightly. The misalignment creates a gravitational torque of magnitude $\tau = mg\xi\sin\theta$, and direction perpendicular to the plane formed by r_m and mg.

"The torque is equal to the change in total angular momentum, but for a rapidly spinning top the total angular momentum is dominated by its spin component. In other words we can take

$$L = I_s\omega_s$$

and in this approximation the torque changes the direction of the angular momentum but not its magnitude.

Caroline: "So the top starts to precess because of the torque?"

Mary: "Exactly. When the motion settles into a quasi-steady state precession at rate Ω about the vertical, the magnitude of the torque, by definition, equals $\Omega L \sin \theta$. If we equate this to $mg\xi\sin\theta$, the result is

$$\Omega = \frac{\tau}{L} = \frac{mg\xi}{\omega_s I_s} \qquad "$$

Caroline: " And I suppose if you use a solid cylinder as an approximation for the top, the precessional speed is $\Omega = \frac{2g\xi}{\omega_s R^2}$ "

Mary: "You've got."

Caroline: "Actually, I don't. I understand the algebra, but what's a steady state?"

Mary: "*In a steady state*, a system's properties remain constant over time, even though processes or changes are still occurring within it. In other words, *the inputs to the system are balanced by the outputs, and there's no net accumulation or depletion of energy, matter, or other quantities.*

"A good example involves the energy from the Sun entering the Earth system—atmosphere, oceans, and land—and the energy leaving it. If greenhouse gases and other anthropogenic emissions weren't entering the atmosphere, the energy entering and leaving would be equal, and the planetary temperature would be in balance when averaged over decades. I'm sure that someday we'll talk about that in detail.

"For now, to return to the topic at hand, a steady state is achieved when the forces and torques acting on the top are balanced, causing it to precess at a constant rate."

Caroline: "You're like Dad. You can't resist the opportunity to pun."

Mary, disregarding Caroline's quip, continued, "With respect to the equation we derived, one thing to notice is that *the angular velocity depends inversely on the square of the radius and directly on the length of the symmetry axis—so the thicker the center of the top, the slower the precession. Another interesting aspect is that as spin slows due to friction, the top precesses more rapidly, even as it becomes less stable.*"

Caroline: "Wow. I hadn't realized that *the width has* such *a strong effect on precession*—and I wouldn't have guessed that *as the top slows down, its precession speeds up!*"

Mary: "Yes, both relations are a bit of a surprise—at least for the uninitiated. But of course, no steady state is permanent. Because the top's steady state lasts for such a short time, I like to use the modifier *quasi*. It's just one of my idiosyncrasies; most people wouldn't bother with it.

"In any case, the steady state doesn't last long because the first law of thermodynamics dictates that energy is conserved—and since friction at the pivot gradually dissipates energy as heat, the mechanical (rotational and spin) energy of the top must decrease. As the spin angular velocity ω_s decreases, the angular momentum $L = I\omega_s$ also decreases. The reduction in L makes the top less stable and increases the tilt angle θ, and over time, the motion becomes increasingly wobbly as the precession becomes less uniform.

"Eventually, the top reaches a critical point at which the spin is too slow to counteract the gravitational torque, and it suffers the same fate as Newton's apple—but it arrives at its final destination in a much more fascinating and dynamic way."

Caroline loved the thought that she now understood something so common yet so mysterious that people are often at a loss for words when asked to explain it. She couldn't resist sharing her new insight with Noreen and some of her other friends. She was, in fact, so excited that she even explained the spinning top to Dr. Kahn—forgetting, for a moment, that he already understood it. But he was gracious, smiling and listening attentively. Then, as some anxiety provoking words left his lips, he pointed to a gyroscope sitting on a shelf among his books. "Now that you understand the top, he said, I'd like you to prepare a report on the gyroscope and present it to the class in two weeks." And then he added with a subtly teasing grin—"I wouldn't want you to get bored,"

Caroline's jaw dropped, wondering what she had gotten herself into. She had heard of gyroscopes, but had never seen one—they weren't everyday objects—and she had no idea how they worked. Or at least she thought she had no idea. Before too long, however, she had recovered her composure and started working her way through the references Dr. Kahn had given her, and building on her understanding of the spinning top.

She soon discovered that the ability of the top to resist gravity and the ability of the gyroscope to keep moving objects such as boats and planes stable, stems from the the generation of a torque perpendicular to the surface reorienting the angular momentum as the top begins to fall. In fact, if the pivot point of a gyroscope is located at the end of its long axis—resting on a support—it is essentially a top. A gyroscope also has a massive disk perpendicular to its spin axis and passing through its center of mass, which is needed to add heft to its moment of inertia.

As Caroline read further, she began to understand why gyroscopes, can stabilize moving objects against random forces, whereas tops cannot. In a typical gyroscope the spin axis is mounted in a so-called gimbal system that allows it to pivot freely about its center of mass—meaning that gravitational torque is essentially absent.

She also felt good as she learned a bit of nautical terminology. She was reminded, though she knew intuitively, that a ship has three degrees of rotational freedom: it can *roll* around its longitudinal axis, which is in the horizontal plane and runs bow to stern passing through the ship's center of mass; it can *pitch* along its transverse axis (bow/stern rising or falling), also in the horizontal plane, passing through the center of mass and perpendicular to the longitudinal axis; and it can *yaw* about a vertical axis perpendicular to the horizontal plane, and passing through the center of mass. What she was unaware of is that ships have 3 more degrees of freedom, which aren't rotational—surge, sway and heave. There's always something yet to be learned.

She sketched some crude diagrams as she thought about the details of these motions. "Suppose there's little or no wind, and the boat moves smoothly with the spin axis aligned with the direction of motion (along the boat's longitudinal axis), when a gust suddenly *yaws* the boat to the right. Then, in keeping with the conservation of angular momentum, instead of simply rotating with the yaw force, the gyroscope resists the torque by precessing—specifically, around the roll axis of the boat." She visualized this by looking at Figure 1, but imagining the spin axis—initially aligned with the boat's motion—pointing into the page.

All that remained to bring her analysis to completion was the algebra, and that was easy—she had already worked through it with the spinning top.

"The angular momentum is given by

L = Iω

where I is the moment of inertia and ω is the spin rate (angular velocity).

A force—whether gravity or a gust of wind—applies a torque,

τ = r × F

where F is the applied force and r is the lever arm from the axis of rotation to the point of application.

Since angular momentum changes in response to torque,

$$\mathbf{dL/dt = \tau,}$$

and this leads to a precession rate

$$\omega_p = \tau/L$$

"So the greater the angular momentum, L, the smaller the precession rate for a given torque. That means more resistance to disturbance. To stabilize a massive object like a boat or a plane, either the moment of inertia I or the spin rate ω—or both—must be large, ensuring that L is large and the system stays steady under external forces."

It took two full weeks for Caroline to complete this assignment, but when she had, she felt the wonderful sense of relief and satisfaction that comes after a rewarding struggle. She now felt fully prepared for Dr Kahn's quizzing, and looked forward to her first teaching experience.

It was a difficult experience and it worked out well, but she was exhausted, and with encouragement by her parents, she was trying to train herself to settle down to a reasonably normal pace.

Now more than a week later, the Angstroms find themselves in their family room just after dinner, about to listen to the news of the day. Caroline's mind, however, is elsewhere. The last few weeks had been so intense that thoughts of those angular momentum mysteries, though beginning to subside, still lingered on.

Caroline:[1] "Now that I know a little about angular momentum and understand the top and the gyroscope, I also have a glimmer of insight into something that must have mystified people since ancient times:

_____

This question and the follow-on between Caroline, who was in her third year of high school, and her mother is based on an exchange that actually occurred, but it involved a boy who was only in the 7th grade. Five years later that boy, Shelley Glashow, after a storied 4 years at the Bronx High School of Science where he stunned fellow students and teachers alike as it became apparent that he knew more chemistry than his chemistry teacher, entered Cornell as a physics major. After graduating he joined Harvard, earning a PhD in theoretical physics, and ten years after that, in 1979, he was awarded the Nobel Prize. Curiosity, indeed! This might be a good place to quote Einstein's comment about his passionate curiosity, but I'll leave that bit of historical research to the reader.

why the Moon doesn't fall to Earth. I guess it's a competition between gravity wanting to pull it in, and angular momentum keeping it moving to the side."

Although Caroline spoke those words in a musing sort of way, half talking to herself, they weren't lost on her mother.

Mary: "You're close! Tie it in with Newton's First Law and inertia. What path would the Moon follow if the Earth's gravity were suddenly switched off?"

Caroline: "OK, I'll give it some thought. But I have another question."

Mary: "Go ahead, but there's not much time — you'll be boarding soon."

Caroline: "Why does the same side of the Moon always face the Earth?"

Mary: "Because the Moon takes almost exactly the same amount of time to spin once on its axis as it does to orbit Earth. The periods are perfectly synchronized to within very small oscillations called librations."

Caroline: "OK, I think I get that. But why are both periods the same?"

Mary: "Great question! It's due to something called tidal locking —and how the periods became locked in synchrony is fascinating, but there's no time to explain that now. Ask your Uncle Zack, I'm sure he'll be able to explain it."

Caroline thanks her mother and says nothing further, appearing slightly anxious as she approaches security.

Airport Security Imaging Is Not a Serious Health Hazard

Caroline: "My science teacher told us about some famous experiments by a biologist named Hermann Muller, who won the Nobel Prize for demonstrating that X-rays induce mutations in fruit flies—and we also learned that there's a lot of evidence showing that mutations can cause cancer. I've heard people on TV news channels complain about full-body scans that display detailed images and about the potential hazards of the radiation used to create those images. I tried to read up on the biological effects of radiation, but I didn't understand much—the articles I found used terms like 'rem' and 'sievert,' and assumed that readers would know their meanings."

Don tried to reassure his daughter, telling her that she didn't need to worry about either privacy or radiation exposure, and that he wouldn't let her fly if he thought either was a problem. But he admitted that he didn't remember much about radiation, and suggested that she ask her uncle about it. Caroline followed her father's advice and, upon arriving in Santa Fe, immediately asked her uncle about the anxiety-provoking imaging booths.

Here's the way the conversation with her uncle would go.

Zack: "'Rem' stands for roentgen equivalent man, but that doesn't give much of a clue about what it means. Let's start with the definition of a roentgen."

Before he could continue, Caroline, with good-natured humor, chimed in, "I'll bet you're going to tell me that a roentgen was named to honor Wilhelm Roentgen, the physicist who discovered X-rays, right?"

Zack smiled. "Yes, exactly. Scientists like to name things after themselves. People in the performing arts like to take bows, do encores, and take more bows. And people in finance…"

Before Zack could finish, Caroline interjected, "People in finance just love to make money and live in big houses—like my mom does," she said with a playful, slightly condescending grin.

Zack looked at Caroline, surprised by her tone. "Caroline, you're being disrespectful and unfair, but I understand your point, and I'm sure you're only half serious. As for wanting to have your 'name in lights,' I think the greatest satisfaction for a musician or someone like your father comes from bringing joy and even ecstasy to an audience by creating or performing the sublime and complex patterns of sound that emerged from the minds of extraordinary individuals who lived hundreds of years ago. The idea that we can connect with members of the human race who are so far removed from us in time and space, but with whom we can share the deepest emotions, is remarkable in itself—not to mention the emotions themselves."

Caroline felt a pang of realization. Her uncle's words had made her see music in a new light—something she had never fully grasped until now. Even her father, a professional musician, had never expressed it in such a way.

Caroline: "It sounds as though science pales in comparison to music and the arts."

Zack: "To a great extent, the arts help make human life worth living. But there's far more to science than most people realize. The world we live in, the origin and evolution of life, the emergence and development of human societies, and the evolution and organization of stars and galaxies—even the origin of matter itself—all of these are utterly mysterious. The search for understanding and the attainment of even a small measure of insight is incredibly gratifying. The problem is that unlike the shared appreciation of artistic creativity, the abstractions of scientific creativity are difficult to grasp, so the joy of sharing that understanding is largely lost."

Stunned by her uncle's words, Caroline felt as though she had been exposed to ideas far beyond what she could have imagined. After regaining her composure, she quickly thanked her uncle and steered the conversation back to their original topic.

Zack: "Yes, indeed, let's get back to the definition of *roentgen (R), which is a measure of* ionization—more specifically, it provides information about *the number of electrons that are knocked free from atoms by electromagnetic radiation* such as X-rays or gamma rays."

Caroline: "I'm sorry, Uncle Zack, but I'm afraid I'm getting a bit lost. In fact, I'm not sure I really understand the difference between ionizing and non-ionizing radiation. I assume *radiation is classified as ionizing if it has enough energy to knock electrons free from atoms*?"

Zack: "Yes, you've got it! If you knock an electron off an atom, you'll create an ion. I don't know what you've learned in school so far, but the energy of an electromagnetic wave is proportional to its frequency. So *high-frequency radiation, such as X-rays and gamma rays, carry a lot of energy and can do a lot of biological damage, whereas low-frequency radiation, such as radio waves and microwaves, are much less energetic and therefore less likely to be damaging.*"

Caroline: "OK, now I get it—reminding me about the relationship between frequency and energy was very helpful."

Zack was just about to continue with his discussion of radiation units when a thought caused him to hesitate.

After a momentary pause, he added, "You know, it's not necessary to go into details about rems. The definitions can be confusing, and the rem is no longer in use. What really matters is how much heat those ions produce in a kilogram of a specific type of matter, and the type of matter matters."

Caroline grinned. "There you go again, Uncle Zack, with your corny wordplay. You and Dad must have inherited the same humor gene, or maybe it's just that you grew up in the same household."

Zack comments on Caroline's allusion to the nature vs. nurture debate (the relative contributions of genes and environment to a trait) and suggests that they discuss it on another occasion. He then continues, unfazed.

"Because of concerns about privacy and safety, and because the concentration of ions is not a suitable surrogate for a biological effect, the rem has been replaced by the gray (Gy), which is expressed in SI units and measures the absorbed dose of ionizing radiation: *a dose of 1 Gy results in the absorption of 1 J of heat per kg of absorbing material.*"

Caroline: "Well, that's a little more intuitive. I learned about heat capacity in my science class: it *takes about 4,200 J to heat 1 kg of water by 1 degree Kelvin*—so now I know that a dose of 1 Gy transfers very little heat to tissue. But that still doesn't answer the central questions: how many grays does an airport scanner deliver, and what are the biological effects of the delivered dose?"

Zack: "Let's start with the fact that *the biological effect of radiation is* not expressed by the Gy. It's *estimated by multiplying the number of grays by an experimentally determined weighting factor, a number that accounts for the type of radiation and the material being irradiated.* The resulting unit is called the *sievert (Sv)*. Of course, it is still in SI units since the weighting factor is dimensionless—it's just a number. So, to understand the biological effects of radiation, you need the definitions of only two units: the gray and the sievert."

Zack takes out some notes and continues.

"With respect to the radiation dose delivered by X-ray airport scanners, it seems that a typical scan delivers a dose in the range of 0.05 to 0.1 microsieverts (μSv), where the Greek letter μ represents $\mu = 10^{-6}$—that is, *one microsievert is one-millionth of a sievert.* To put these numbers in perspective, *a typical chest X-ray delivers a dose of approximately 10 μSv;* natural *background radiation at sea level, from cosmic rays and the natural radioactivity of the planet, exposes us to about 3,000 μSv per year.* Radiation exposure when flying across the U.S. is in the range of 20–30 μSv.

"Although there is an enormous amount of evidence that some mutations can cause cancer, the comparisons I just summarized indicate that radiation delivered by X-ray backscatter scanners used at airports is considerably less than background radiation.

"Having said all that, I should add that these calculations don't matter much because U.S. airports have abandoned X-ray scans in favor of microwave scans—what they euphemistically refer to as millimeter-wave scans, because authorities and congressmen know that some constituents are concerned about the potential harm of microwaves. Microwaves, however, are non-ionizing radiation and there is no evidence that they mutate DNA. Nor is there evidence that, at the dosages used, the heat they produce is damaging—and the heat isn't cumulative, unlike the damage caused by X-rays."

Regarding privacy, Zack notes, "Congress has prohibited the display of detailed images and mandated that metal and other objects be shown on a generic body outline, represented in a cartoon-like format instead of the person's detailed appearance."

Caroline: "OK, Uncle Zack, I get the idea—I'll feel less anxious on my flight home than I did on my trip to Santa Fe."

That's how it would go, but it hadn't yet happened, and now Caroline was about to pass through security and was still uneasy—but that was part of life in the 21st Century.

Conversation 4

Color, Rainbows and Mirages

As Caroline travels to Los Alamos to visit her father's brother, she muses about Davey Wexler, the boy in Judy Blume's Tiger Eyes who made a similar journey. Uncle Zack, who hadn't seen Caroline in some time, was struck by how much she had grown and commented on her resemblance to both her father, who has Swedish roots, and her mother, who has a Greek and Sicilian heritage. As they talked, Caroline asked a series of questions about the origins of different skin colors, prompting Zack to reflect on the endless nature of curiosity and how long one can keep asking "why". As the visit came to an end, and Zack drove her to the airport, the awesome sight of a complete rainbow sparked a conversation about the science of color and led Caroline to wonder whether rainbows are just mirages.

Mary and Don embraced Caroline warmly, their eyes misted with a blend of pride and a subtle, unspoken melancholy as they felt, perhaps for the first time, the bittersweet recognition that her childhood was drawing to a close. On the other side of the security scanners, an airport attendant would escort her to the gate, where she would board her flight to Dallas before transferring to a connection to Santa Fe. Although she was slightly anxious, it was tempered by an unfamiliar and oddly comforting sensation—she was on her own, and beneath the apprehension, a new sense of confidence was taking root. For the first time, Caroline felt herself crossing into the realm of adulthood.

The transfer in Dallas, assisted by airline staff, was smooth. As soon as she was seated on her flight, Caroline called her parents, her voice steady, before settling in with her book, *Tiger Eyes*, by Judy Blume. The story of 15-year-old Davey Wexler, who spent months with his aunt and uncle in Los Alamos after a tragedy, resonated deeply with Caroline. The parallels were uncanny: they were the same age, both from the New York area, and both were visiting a close relative in Los Alamos.

The flight was brief, around two hours, and her Uncle Zack was waiting for her at the gate. Although they had seen each other only a year earlier when Zack stopped off in New York on his way to Washington, D.C., Caroline had grown several inches.

Zack: "You've grown into quite an attractive young lady: tall and slender, with straight dark hair like your mother's."

Caroline: "But I've got my father's blue eyes."

Zack: "Yes, you do, but that's about it. You even have your mother's olive complexion. Your father's Swedish background is hardly noticeable, at least in terms of superficial physical characteristics."

Caroline's paternal grandparents were both born in Sweden, but her mother's mother was born in Greece, and her mother's father was born in Sicily.

Zack explained that her aunt's name, Zoe, is the Greek word for life. "I first heard the name when I was in high school, in a lovely poem by Lord Byron, *Maid of Athens*. Byron lived in Greece for a brief time, fighting in the Greek War of Independence against the Ottoman Empire. The story goes that while he was there, he fell in love with a beautiful Greek maiden. The poem contains the line Ζωή μου, σᾶς ἀγαπῶ—'My life, I love you.'"

Caroline was barely listening. "That's interesting, Uncle Zack, but actually I had a question. Why are people from Southern Europe, where Mom's parents are from, darker than people from Northern Europe?"

The simple question took Zack by surprise; all he could manage was a trite, "I'm glad you're still asking questions."

Caroline: "It seems like an obvious question…"

Zack: "Yes, in a way, it is. But sometimes—often, in fact—when we see things every day, we tend to accept them without asking why."

Zack shifted gears with what seemed like a non-sequitur. "Have you ever heard of the *Encyclopedia Britannica*?"

Caroline: "No, I haven't."

Zack: "It was a set of a few dozen large volumes in which you could find information on almost anything. One of their very clever advertisements had a young boy asking his father why the sky is blue. The father, who had never thought of the question, couldn't answer."

Caroline's eyes lit up. "OK, I get it. But another question just occurred to me: Why do I look dark like my mother instead of fair like my father, or somewhere in between? And since you brought up the subject, why is the sky blue?"

Zack's smile faded as he considered her questions. "Those are difficult questions—try the *Encyclopedia Britannica*."

Caroline: "Uncle Zack, your sense of humor is like your brother's."

Understanding Why a Blue Sky Turns Reddish Orange in the Evening Involves Both Physiology and Physics

Zack: "OK, let's start with why the sky is blue.

"There are two key concepts underlying an explanation of the color of the daylight sky and why it changes from blue to reddish orange as the Sun sets. The first, demonstrated by Isaac Newton, is that ordinary *white light consists of a spectrum of colors, ranging from red, which has the longest wavelength, 650–700 nanometers (nm), to blue (450–495 nm), and violet (300–450 nm),* which have the shortest wavelengths."

Caroline: "I learned a lot about units, but I never came across the word nanometer."

Zack: "Nano means 10^{-9}."

Caroline: "So *a nanometer is 10^{-9} meters?*"

Zack confirmed and continued his explanation:

"The second concept, demonstrated: mathematically by Lord Rayleigh (after whom it is named), is that *when visible light scatters off atmospheric molecules, especially nitrogen (N_2) and oxygen (O_2), its intensity varies inversely as the fourth power of its wavelength.* The Rayleigh relation tells us that if one were to look directly at the Sun (which of course should never be done without protection), it would appear white or yellow because all wavelengths would reach the observer. However, scattered light from other parts of the sky (not from the direct Sun's path) is predominantly blue due to the inverse-fourth power dependence of Rayleigh scattering.

"*In the evening, when the Sun is low in the sky, the distance that light travels through the; atmosphere is considerably greater than it is during the day so that blue light is scattered out of the direct path before reaching the observer, leaving red and orange hues which travel the long, low path from the Sun to an observer with relatively little scatter.*

"*Another factor is the detector*—the human eye, which has cell populations called cones that are differentially sensitive to wavelengths in the red, green, and blue portions of the electromagnetic spectrum. The *cone cells sensitive to blue wavelengths are in a distinct minority, comprising only 2–7% of the population, while those with maximal sensitivity to red are at 64%.*"

Caroline: "What would the color of the sky be if the sizes of the red and blue populations were reversed?"

Zack: "If the cone cells sensitive to blue were in the majority, and those sensitive to red were a small minority, the evening sky would have a more bluish hue, while the intensities of the reds and oranges would be substantially diminished.

"These effects are amplified somewhat because, at sunset, the Sun is already below the horizon, and only refracted light reaches an observer. Consequently, light travels an even longer distance than it would if the Sun were still above the horizon."

Caroline: "This is all new to me and interesting. I never thought about why the sky looks blue during the day and reddish-orange in the evening. The explanation is fascinating and understandable, even though it's a bit complicated. Now, what's the answer to my other question?"

Caroline's Mother is Darker than her Father Because her Skin has a Higher Concentration of Light-absorbing Molecules

Zack chuckled. "The simple answer to your question—why *some people are darker than others— is that people with darker skin have more color pigments than those with lighter skin.*"

Caroline: "What are pigments?"

Zack: "*A pigment is a molecule that absorbs specific frequencies of light.* An example you might be familiar with is chlorophyll, which is essential to photosynthesis. It's responsible for the green color of leaves during their growing season because it preferentially reflects green wavelengths while absorbing red and blue wavelengths.

"In humans, *skin color is mainly determined by molecules called melanin.* Generally speaking, the more melanin, the darker the skin."

The Search for Understanding Can Lead to an Infinite Regress of Causes and Whys

Caroline tilted her head. "You've explained how skin color is determined, but what I really wanted to know is why people in Northern Europe developed small amounts of melanin, or none at all, making them relatively fair, while those in Southern Europe have more, making them relatively dark."

Zack's eyes brightened with understanding. "That's a nice and very important distinction you're making—I mean between how and why. OK, let me try again. Around 60,000 years ago, during the last glacial period when ice sheets covered large parts of the Northern Hemisphere…"

Caroline's eyes widened. "Ice sheets covered a large portion of the Northern Hemisphere? Was New York buried in ice?"

Zack laughed softly. "Yes, *over the last million years or so our planet has regularly passed through glacial periods—so-called ice ages—that last about 80,000 years, and interglacial periods, like the one we're in now, that last approximately 20,000 years.* But let's save that discussion for another time or I'll never get back to your questions about color.

"As I was about to say, around 60,000 years ago the ancestors of all the people who now inhabit our continents lived in the tropical deserts of northeastern Africa. They were all the same color, probably brown."

Then Zack surprised Caroline by asking what she knew about light, a question that at first seemed unrelated to their conversation.

Caroline: "I learned a little in school, and Dad also taught me a few things. I know *light can be thought of as a wave that travels at 186,000 miles per second, and that most of the light around us—like ultraviolet light and infrared—can't be detected by our eyes*."

Zack: "I see you do know something about light—no pun intended. Now, here's the key. Ultraviolet light, which has a higher frequency and more energy than visible light, can be damaging to a person's skin and eyes. It turns out that *one type of melanin provides protection against UV and also gives skin a brownish color*."

Caroline: "Oh, it's about light again, I see. No pun intended."

Zack: "I see you're a bit of a smart aleck, no pun intended."

He smiled slightly and continued. "Those of our ancestors who lived in warmer climates where the Sun is strong, and who happened to have high concentrations of this particular kind of melanin were more likely to reach reproductive age without severe UV-related damage—such as skin cancer or the degradation of certain B vitamins that are essential for cell growth—making them more likely to have healthy offspring. As the generations went by, those who lacked the protective melanin became a smaller and smaller percentage of the population. Those who lived in the cooler climates didn't need the extra protection, so brown melanin offered no advantage.

"Another possibility involves the interplay between UV, melanin and vitamin D synthesis. Since regions further from the equator receive less intense ultraviolet (UV) radiation, and vitamin D synthesis relies on UV radiation, the efficiency of vitamin D synthesis decreases as distance from the equator increases. What does that bring to mind about a relationship to skin color?"

Caroline: "I don't know..." (She pauses in mild frustration, then suddenly lights up.) "Oh, I think I know—lighter skin enhances UV absorption so it evolved as an adaptation."

Zack: "Yes, that could ensure sufficient vitamin D production for bone health and immune function—and even prolongation of life."

Caroline frowned. "That last possibility doesn't sound right. Women have children at a relatively young age, so how can living beyond the age at which reproduction occurs be advantageous if it doesn't affect their ability to reproduce?"

Zack nodded thoughtfully. "You're right. Maybe it's the advantage for men that matters. I'll have to do some research. In fact, why don't we both look into it?"

Caroline: "OK, sounds interesting. Isn't what you just described called *natural selection*?"

Zack: "Yes, that's the general idea."

Caroline hesitated. "I'm sorry, Uncle Zack, I have another question. I understand that melanin, which is responsible for brown skin, offers no advantage in cooler climates, but why don't we see both fair and dark-skinned people there? Is dark skin a disadvantage in cooler climates?"

Zack was stumped once again—after all, he wasn't a biologist, and all he could do was speculate. "Actually, I don't know—that's a good question. I can imagine reasons why producing melanin when it's not needed could be disadvantageous, since there's an energy cost to its synthesis. Or it might provide some advantage unrelated to protection against skin damage. It's also possible that fair skin offers an advantage. I don't know if there are definitive conclusions, but your Aunt Zoe is in a better position than I am to explain the most recent hypotheses. By the way, how is my favorite sister-in-law?"

Caroline smirked. "She's your only sister-in-law, so she has to be your favorite."

Zack raised an eyebrow. "She could also be my least favorite."

Caroline rolled her eyes, ignoring her uncle's last remark. "She's very busy as usual."

Zack smiled. "I'm sure her position as an associate professor keeps her fully occupied. There's probably no other profession having the productivity demands of a faculty member at a research university."

Caroline: "Really? Mom says that Aunt Zoe generally teaches only one course a year—and she also has the summer off. Sounds pretty easy to me."

Her Uncle Zack, who feels that most people don't fully understand what university life entails, began to elaborate. "Maybe I can best give you an idea of what's involved in being *a faculty member at a research university* by using the language of motion pictures: a professor *is a writer, producer, director, creator, actor, and publicist—and that's only part of it.*"

Caroline: "What? I don't get it."

Zack: "Producer, because she has to raise money to support a research team; writer, because she has to write research proposals to obtain funding and publish the results of her research; director, because she must guide and manage a research team; publicist, because she presents findings to her colleagues at meetings all over the world; creator, because she asks questions and contributes to finding answers. In addition, of course, she's a teacher—not just because she teaches a formal course once a year, but because managing a lab with 12–15 members provides an environment for learning how to think creatively and discover what no one knew before.

"Your aunt also serves on departmental committees, university committees, and federal advisory boards. Additionally, many faculty members, especially in engineering and science, who conduct research with immediate societal implications, start their own companies."

He continued as Caroline listened dumbfounded. "That's what it takes for a university to drive creation and change. Universities deepen our understanding of the world around us, produce new knowledge, and then pass that understanding on to the next generation. And your Aunt Zoe definitely doesn't have the summer off. Research is a full-time occupation."

Caroline: "OK, OK, Uncle Zack, I get it," she finally blurted out, feeling as though she had just been pummeled by a sales pitch.

Zack: "I guess I have strong feelings about universities. Anyway, tell your aunt I said hello," he added with a slight sigh of exasperation.

Caroline: "I will."

Caroline and her uncle continued walking toward the parking lot, but in silence, as she tried to process everything she had heard during the last ten minutes. A subtle sensation lingered in her mind—an inkling that her uncle's feelings for her aunt might be special, different from the typical in-law relationship. But she had no real frame of reference for how in-laws usually felt about one another, and the thought quickly passed.

Caroline was so focused that she didn't notice one of the most unique and charming airports in the country. The space featured wood beams across the ceilings, floor tiles patterned with designs characteristic of the Southwest, soft leather furniture, and architecture marked by gentle curves and reddish clay adobe construction. But she would soon see that and so much more during her visit: the otherworldly, massive rock formations and mesas, the vast open terrain, the majestic mountains that glowed red at dusk, and, of course, the unique architectural style that captured the character and charm of the Old West. It would be unlike anything she had ever encountered. The sights would take her breath away, causing her eyes to close reflexively, as if to retain the images. And each morning when she opened her eyes, that feeling would return.

If Happy Little Bluebirds fly

The week flew by. Zack took a few days off from his responsibilities at the Lab to show his niece around Santa Fe and Los Alamos. They visited Bandelier National Park, where Caroline climbed large rocks to reach the dwellings once inhabited by Pueblo Indians. They drove through the Jemez Mountains, where she learned about the Valles Caldera, the remnant of a super-volcano that last erupted more than 50,000 years ago—long before *Homo sapiens* inhabited North America. She even imagined seeing the canyon described in Judy Blume's novel, where Tiger met Wolf.

Caroline discovered that both Los Alamos and Santa Fe are situated more than a mile and a quarter above sea level, which results in lower atmospheric pressure and cooler air temperatures. This altitude also stimulates the production of additional red blood cells to compensate for the lower concentration of atmospheric oxygen. And—most importantly for a New Yorker—the reduced pressure presents serious challenges for bagel bakers. Caroline had countless questions, some of which her uncle answered, but most were added to her father's list for those rainy Saturdays.

Now, the visit was almost over. They would drive to Albuquerque International Airport early the next morning to take advantage of better flight options. They'd had huevos rancheros for breakfast, and Caroline was still trying to decide whether she preferred bagels and lox or the wonderfully complex mixture of eggs, tomatoes, and spices she had just finished with uncharacteristic gusto.

Figure 1. A monochromatic rainbow often occurs after a rain shower just as the Sun is rising.

A Rainbow is Caused by the Refraction, Reflection, and Dispersion of Light Within Water Droplets

The Sun had just risen, and the air was clear, moist, and fresh after an unusual early morning shower as Zack drove southwest along a stretch of Interstate 75.

Zack, with excitement in his voice: "Caroline, look out the rearview window."

Caroline turned and, for a moment, felt as though she were in the land of Oz, encountering a sight she had previously seen mainly in movies or storybooks.

Caroline: "It's awesome! I've seen rainbows in movies, but rarely in real life."

Zack: "This one's relatively simple—it's *monochromatic*. Rainbows are quite common here in the open Southwest, especially at dawn and dusk, and right after rain showers."

Caroline: "Why do we see so few in New York City?"

Zack: "I'm not sure; I suspect pollution plays a role—dust and other particles in the atmosphere could interfere, but that's probably a secondary factor. More likely it's the density and height of buildings, which limits views of an open sky."

When they were about halfway to Santa Fe, Caroline saw something even more amazing—stretching across the stark open terrain was a sight straight out of a fairy tale: a complete rainbow.

Zack could see the question forming on her lips and half-hoped she wouldn't ask, but it was as though the question was fated to be spoken.

Caroline: "Uncle Zack, what are rainbows? Can we drive to them? Will they be disrupted if we drive through them?"

Zack: "*A rainbow is not a physical object*—so what do you think the answers to your other two questions are?"

Caroline: "Well, if it's not a physical object, I guess the answer is no to both. But what causes it?"

Zack: "The short answer is that it's caused by the way sunlight interacts with water droplets dispersed in the atmosphere. But explaining exactly how that interaction produces the colors and characteristics of a rainbow is difficult without drawings, and you'd also need to know a bit about optics. Ask your father to add it to his Saturday list."

Caroline: "That's disappointing."

Zack: "I know. Sometimes, you have to study for a while before you can understand a phenomenon. True understanding often takes patience and discipline—you might as well start getting used to it."

A Mirage is Caused by the Refraction of Light in Layers of Air with Varying Temperatures and Densities

Caroline persisted in trying to draw explanations from her uncle, asking whether a rainbow is just a mirage.

The question pushed Zack into territory he was trying to avoid and compelled him to think about how to explain the ideas of physical optics to his niece, who had only studied elementary physics.

Zack: "That's a good question, but the answer is no. *Rainbows and mirages are both optical phenomena, but the physics that explains them is entirely different.* In fact, the explanation of a *mirage involves both physics and biology.* When you see a mirage, you're seeing refracted light that might or might not have originated from a physical object. *People in the hot desert sand who think they're seeing a body of water in the distance are actually seeing the sky and misinterpreting what they're seeing.*"

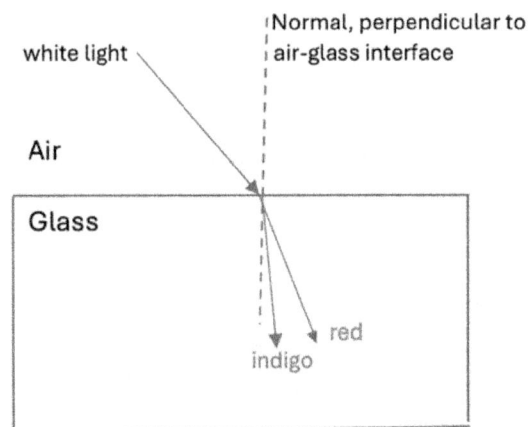

Figure 2. When white light enters a medium with a higher refractive index, as it does when it passes from air to glass, it slows due to interactions with the silicon dioxide molecules of which glass is composed, and bends toward the normal. Since the higher frequencies interact more strongly than the lower frequencies, refraction will be greatest toward the indigo/blue end of the visible spectrum. Consequently refraction will be accompanied by dispersion of white light into its component colors.

Zack was relieved when Caroline told him she remembered learning about refraction in her physics course.

Caroline: "As I recall, *refraction is just the bending of light when its speed changes as the result of passing between media with different properties.* So when light passes from air into glass, or vice versa, it bends because glass and air have different optical properties."

Zack: "Exactly. Since air is less dense than glass, optically as well as physically, the speed of light decreases, and it bends toward the normal when it enters glass at an angle."

Caroline: "I'm not sure I know the difference between physical density and optical density. I know physical density is the amount of mass per unit volume but what's the optical density? Is it related to how much light can pass through an object?"

Zack: "Yes, you've almost got it, but not quite. *Optical density is more accurately described as how much a material slows light relative to a vacuum*, expressed as the refractive index (n). A material can have a high refractive index but still be transparent."

Caroline: "Great, I have just two questions"

Zack, sensing the limited time before Caroline's flight, feels himself about to dive into territory he was hesitant to explore, but responds light-heartedly, saying he's glad there are only two.

Caroline: *"Why does the speed of light increase as the optical density of a medium decreases, and why does light bend when it passes into a less dense medium?"*

Zack: "The thing to keep in mind is that as light travels through any medium—anything that isn't a vacuum—it interacts with the molecules that make up that medium. What do you think the consequences of that are?"

Caroline responds with an uncertain tone: "Does the light slow down?"

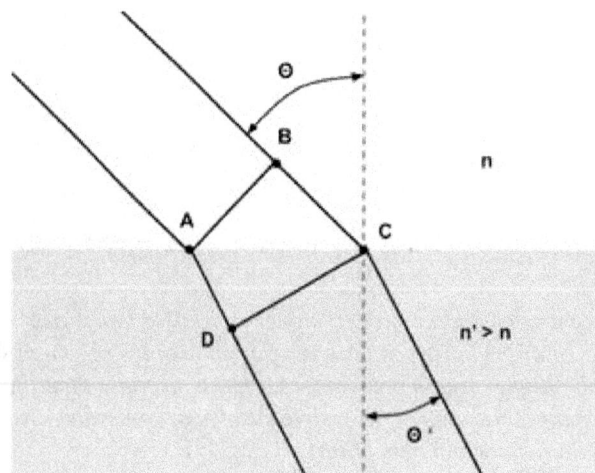

Figure 3. When a light wave enters a medium with index of refraction n' > n at an oblique angle the portion of the wave that has crossed the interface slows down relative to the segment that hasn't crossed. Consequently it will bend toward the normal because it will not have traveled as far as it would have if it maintained the same velocity.

Zack: "Yes, you're on the right track, and it will make more sense as we continue."

Caroline: "OK, I think I see the answer—the denser the material, the greater the number of collisions, and the slower the speed. But why does light bend?"

Zack: "That's a bit more complicated, but actually not too hard to understand. You know that light can be thought of as a wave, so try to picture it entering a denser medium at an oblique angle— that is, at an angle that's not perpendicular to the interface."

Caroline pulls out her phone and finds a diagram illustrating refraction that essentially provides the answer.

Caroline: "When the leading edge of the light enters the higher optical density medium, it slows down relative to the part of the wave that's still in the lower-density medium. As a result, the light bends toward a line that's perpendicular to the interface."

Zack: "Great, you've got it."

Caroline: "I've got just two questions."

Zack, his response somewhere between frustration and playful repartee: "I thought we just answered two questions."

Caroline: "We did, but I have two more. We were starting to talk about mirages, and of course that immediately conjures up an image of the desert and of hot sand. So my question is, why does sand get so much hotter than air? And when I think of scorching hot sand I'm also reminded of the beach; though the air is a bit uncomfortable, the sand is so hot that walking on it is nearly impossible."

Zack, with no small amount of anxiety: "And the second question?"

Caroline: "How does refraction cause a mirage?"

Zack: "The first question can be answered by physics alone, but the second is a bit trickier because it also involves biology."

Caroline: "Go ahead. I'm ready."

Zack: "The main reason that sand can get much hotter than the surrounding air is that it absorbs sunlight directly and efficiently. Air, on the other hand, is largely transparent to visible sunlight and is heated indirectly by conduction from the ground and re-radiation from the sand. Additionally, the specific heat of sand is much lower than that of air, so sand heats up and cools down much more quickly than air does. *Air is also a poor conductor of heat*, so it doesn't absorb and retain heat as efficiently as sand."

Caroline: "Why is that?"

Zack: "I can see we're in danger of spiraling into an infinite set of questions. Let me make one more statement, and that will have to suffice for now. The answer is that sand has a much more tightly packed structure than air. The vibrational motions and mass of its molecules enable it to store energy more effectively than air does. The amount of heat required to raise the temperature of one gram of a substance by 1° Kelvin is so important that it has a name. Do you remember learning about that?"

Caroline: "Yes, it's called the *specific heat, which, as I remember, is the heat capacity per unit mass.*"

Caroline continues: "OK, so specific heat has to do with the details of atomic and molecular structure. That's not completely satisfying, but it's better than nothing. At some point, I'd like to get back to it, but for now what I'm most curious about is what all this has to do with mirages."

Zack: "Let's begin with how light behaves when passing through air layers of different temperatures, and therefore different refractive indices. As light from the Sun travels toward the Earth, it generally moves from air at higher altitudes where it is cooler and at lower pressure, to air near the ground where it is warmer and at higher pressure. What's most important is the temperature difference near the ground. On a hot day, the ground heats the air in immediate contact with it, causing a thermal gradient and therefore a gradient in air density and refractive index. As a result light traveling from above will bend away from the normal as it moves from higher to lower refractive index. This process causes the light to follow a curved path, eventually bending upward toward an observer.

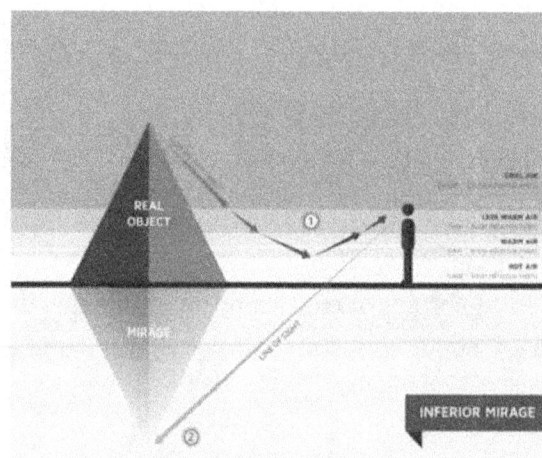

Figure 4. Zack directed Caroline to this web site illustrating a mirage that looks as though it's below ground. As shown in this figure, a mirage can be refracted light from a physical object, or it can be light refracted from the desert sky that looks like a pool of water but is not a physical object.

"Here's the key: our brains assume light travels in straight lines. So when the refracted light reaches the observer's eyes, it appears to come from the ground instead of the sky, creating the illusion of a shiny, water-like surface."

Caroline: "Oh, so the 'water' is actually an image of some portion of the sky being bent upward?"

Zack: "Exactly. The light that's bent upward often has a bright, bluish appearance, which resembles a reflection of the sky on the surface of water."

Zack continues: "The mirage of water in a desert has one thing in common with rainbows: neither of them represents a physical object. However, some mirages do represent physical objects."

To help Caroline visualize what he's saying, he directs her to a website with an illustration of a mirage that looks as though it's below ground.

Caroline: "Can refraction also explain why a car on a hot dry road looks as though it's on a wet road?"

Zack: "Yes, it's the same type of refractive mirage."

Caroline: "The road also looks like it's shimmering — does that have anything to do with the mirage?"

Zack: "Shimmering is caused by small-scale turbulence in the hot air near the ground. Uneven heating creates ripples in the refractive index, which distort the light paths slightly, making the "mirage" shimmer. It mimics the way sunlight might reflect off water, reinforcing the illusion— but it's not the main cause of why we see the mirage."

Caroline: "So, mirages are created by a mix of physics and how our brains process the information?"

Zack: "Exactly. The physics of light bending, combined with the brain's tendency to interpret light as traveling in straight lines, creates this fascinating illusion."

Caroline: "I understand—and by the way, that really didn't take much patience and discipline."

Zack, after a slight grin acknowledging Caroline's dry wit: "It looks like your plane just arrived. I'll walk with you to the gate."

Caroline: "I learned so much about refraction and the way light can trick our visual system that I should be able to understand at least something about *rainbows*. Can you at least give me a one minute overview?"

Zack: "OK, here's a very quick 10,000 foot view. *When sunlight enters a falling water droplet, it is refracted and dispersed into its component colors.*"

Caroline: "Dispersed?"

Zack: "I'm sure you remember— I mentioned dispersion briefly when explaining what happens when light moves from air into glass."

Caroline: "Yes, thanks for the jog"

Zack: "Great because dispersion is key. *Inside the droplet, the dispersed light is reflected by the surface and then exits, undergoing a second refraction. This process further separates the colors, producing the distinct bands of a rainbow.* In a primary rainbow, red light appears on the outer edge and violet on the inner edge. A secondary rainbow forms through two internal reflections within the droplets, reversing the color order and making it appear dimmer than the primary rainbow. It's important to realize that this description leaves out quite a bit and you'll have to do some research on your own to develop a deeper understanding. For example, to observe a rainbow, you must be positioned with your back to the Sun, looking at water droplets at a specific angle relative to the sunlight."

"Well, that will keep me busy! Uncle Zack", Caroline says with admiration, "this has really been fantastic. Thank you so much, it's been awesome seeing and learning things I never even imagined—I'll never forget this trip."

Caroline, boards the plane, sinks into her seat, closes her eyes, and dreams she's with her friend Davey, exploring the limitless expanse of the Valles Caldera.

Conversation 5

Light: We See so Little

Caroline is intrigued by a simple mystery of technology: Why do TV remote controls need to be pointed at the screen, while garage door openers work from a distance and can even function through obstructions? She and her best friend, Noreen, ask Noreen's father — an electrician with a good working knowledge of everyday technologies. Noreen's father begins by explaining that human vision can detect only a tiny slice of the vast
spectrum of light that surrounds us. This spectrum includes x-rays, radio waves, visible light, microwaves, and countless other forms of electromagnetic radiation — all distinguished only by their frequency. He emphasizes that in a vacuum all electromagnetic waves travel at the astonishing speed of approximately 186,000 miles per second, the maximum possible rate at which any signal can move through space. The signals from garage door openers are radio waves that can travel through walls and other obstacles without a direct line of sight, while TV remotes use infrared waves, which require a direct line-of-sight to the TV.

Despite his explanation, Caroline wonders why this is so, and she and Noreen are left wondering why pointing is necessary for TV remotes but not for garage door openers. Caroline feels she is missing something fundamental and turns to her father, hoping he can explain the conceptual basis of electromagnetic radiation. But while she understands the concepts described by her father, her curiosity about electromagnetic radiation deepens — and she wants to know more about what it is and where it comes from.

This curiosity leads her to her Uncle Zack, who shares the story of two brilliant minds — one a genius with no formal education, and the other the leading theoretical physicist of his time who revolutionized our understanding of electricity and magnetism and the intricate connection between them. Their work laid the foundations for modern science and sparked further exploration into how different forms of energy, such as electrical, mechanical, and chemical, can be transformed into one another. Her parents emphasize that achieving an efficient and clean transformation between these energy forms is one of the most significant challenges facing modern civilization.

Although Caroline feels she has learned a great deal, she notices that each answer she receives only leads to more questions. And so, with a mix of awe and amusement, she finally asks the biggest question of all: Will the "whys" ever end?

We're Flooded by Light We Don't See

It's another hectic morning at the Angstrom's, and today Caroline isn't feeling well. Mary Angstrom tells her husband that she took Caroline's temperature, and it was slightly over 100 — "unusually high for this time of day."

Mary Angstrom, looking at Caroline, says, "Stay home and rest, and do some reading if you're feeling up to it. Lucy will come by around noon to straighten up, and she'll also make you lunch if you're hungry."

Caroline thinks to herself, "This is great! I get to stay home from school, and I really feel OK, but I'm not going to tell my mother."

"Hey, Mom, I have a question: You said my temperature is unusually high for this time of day— does my temperature change throughout the day?"

Mary: "That's a good question—and the answer is yes, body temperature does vary during the day. But your question just made me realize that I don't really understand why it changes. Let's try to figure it out together during one of our rainy Saturday sessions. Right now, though, I'm running a little late for a meeting with one of my financial partners."

Caroline, who recently turned fifteen, hasn't yet begun to think much about money—not even a little, which seems surprising given her endless curiosity. A decade will pass before money enters her consciousness, and when it does, she'll see it very differently than her mother does.

Her mother hugs her, rushes out the door, and her father follows, explaining that he has to rehearse for a performance later that evening.

Caroline reads a short story on her iPad but soon becomes bored and decides to watch TV— something her parents usually discourage. But no one will know, and with just a twinge of guilt, she grabs the remote, presses the power button... and nothing happens. She keeps trying without success, until finally, by chance, she points the remote toward the power box as she presses the button, and the TV turns on. She wonders why TV remote controls have to be pointed at the screen, but garage door openers work even when the door isn't in sight.

"That's strange, Mom and Dad never point the garage door remote, which they don't even hold— it stays on the visor of our car—but the door opens every time. Sometimes the car hasn't even entered the driveway yet, and you can't even see the garage door because of all the trees in front of our house. How can I ask them why the TV remote has to be pointed at the television, while the garage door remote can be pointed any which way, without telling them I was watching TV?"

A few days later, on a Saturday morning, Caroline is with her best friend, Noreen, who lives in a modest middle-income apartment building a few blocks away. Caroline expresses her bewilderment about the two remotes, hoping Noreen might have an explanation, but Noreen, too, has no idea why the TV remote has to be pointed while the garage door remote does not.

Noreen suggests they ask her dad, Joe, who is outside in the parking lot, installing a new radio in the family car. Caroline admires Mr. Lewes, an electrician who never graduated from college but is very intelligent and knowledgeable about everyday technology.

Mr. Lewes begins by reminding Noreen and Caroline that "the Sun's energy arrives here on Earth in the form of electromagnetic radiation, most of which is invisible light. You might think the term visible light is redundant, but the truth is we can detect only a small part of the Sun's radiant energy with our eyes, as visible light, and another part with our skin, as heat—infrared radiation. But *I like to call all forms of electromagnetic radiation light*." He adds, "Unlike pressure waves in the air that we detect as sound, light travels through what appears to be empty space."

Mr. Lewes taps on the Google app on his phone and types "electromagnetic radiation" to find some examples: "X-rays: frequencies between 30 thousand trillion Hz (30 Peta Hz) and 30 million trillion Hz (30 Exa Hz); visible light, the light we can see: frequencies ranging from 400 trillion to 750 trillion Hz (400–750 Tera Hz); infrared, from 300 billion (300 Giga Hz) to 400 Tera Hz; and radio waves, less than 3 Giga Hz (GHz)—and that's just the tip of the iceberg."

Caroline's head is spinning and she's talking to herself: "Peta, Exa, Giga, Tera—scientists are really into Greek. I'll have to remember to talk to Aunt Zoe about this."

As Mr. Lewes continues, Caroline listens and at the same time her thoughts swirl: *"X-rays, visible light, radio waves, microwaves! I've heard of all of them before, but I never fully grasped the idea that they're all the same except for frequency, and that our senses respond to an insignificant fraction of the frequency spectrum."*

She marvels at the idea that electromagnetic radiation surrounds us, yet our senses can detect only a tiny sliver of it. "So much of the world is invisible: first it was air molecules, now it's all kinds of electromagnetic radiation—and yet scientists have discovered that molecules and radiation are as real as I am. How sad it is that we can see so little of the world around us; but how remarkable that we can discover what our senses can't detect." Then, in a moment of insight, she realizes that we do detect all kinds of radiation—not through our senses, but through the parts of our brains that enable us to reason and create.

Although she still doesn't fully understand how this relates to her original question, Caroline finds the information fascinating, and it's beginning to change her view of the world. Her thoughts are abruptly interrupted when Mr. Lewes adds, "...*they all travel at the same remarkable speed, a speed that's almost too fast to grasp—186,000 miles per second.*"

She had also heard that before, but now in an attempt to put it in perspective, she pulled out the calculator her father had given her and quickly calculated that light takes just over 8 minutes to travel the 93 million miles from the Sun to Earth. Half in disbelief, she recalculated, but there it is again: 8.3 minutes.

Mr. Lewes, impressed by her curiosity, adds something that she hadn't heard previously: *"Not only does light travel with incredible speed, nothing travels faster."*

He had once read in a popular science magazine that an absolute speed limit does more than reveal a simple fact; it lays the groundwork for Einstein's theory of special relativity.

Mr. Lewes: "I really wish I understood relativity, the little I've heard sounds fascinating. Maybe the two of you will be able to explain it to me someday. For now, though, I can at least tell you about garage doors and TVs."

Mr. Lewes explains that the garage door remote transmits radio signals to a device that controls the door's motor. "These are the same invisible signals transmitted from a broadcasting studio to your radio, allowing you to hear music and news."

Noreen asks, "Then why can't anyone with a radio transmitter open our garage door?"

Her father replies that the signal is encoded—it's broken into packets, and there are essentially an unlimited number of ways to do that.

He adds, "By the way, don't confuse what comes out of a radio with radio waves. The radio receives radio waves and converts them into sound waves, which is what you hear."

Caroline sighs slightly. "Many of the words connected with radio are a little confusing. I've heard reporters refer to broadcasts over the radio as 'airwaves,' even though radio waves have nothing to do with air. They're not like sound waves, which really are airwaves. And it's strange that they're called 'radio waves' when a radio emits sound waves."

Mr. Lewes nods. "Yes, the English language isn't nearly as structured as a scientific vocabulary. But that flexibility adds power to the language, even at the cost of occasional ambiguity."

He then continues, "… in contrast to the garage door remote, your TV remote transmits near infrared radiation (IR) whose wavelength is much shorter than that of radio waves."

Everyday Devices Such as TV Remote Controls and Garage Door Openers are Controlled by Light With Vastly Different Frequencies

Noreen, wanting to be sure she understands, summarizes what her father has explained: *"So, the garage door remote and the TV remote use different signals—radio waves and infrared waves—and the radio waves have a much lower frequency than the infrared waves?"*

Noreen's father: "Yes, for our TV to receive a signal from an infrared controller, you have to point the remote along your line of sight toward the device it's controlling, such as the cable TV receiver."

As Mr. Lewes finishes installing the car radio, he tells the girls that he enjoyed their conversation and reminds them that infrared and radio waves are just two of the many types of electromagnetic energy present in our daily lives. "Engineers have invented methods to produce, control, manipulate, and use all frequencies of radiation in countless ways—all of which make modern civilization possible."

Caroline and Noreen understood the words but they were still a little perplexed. They learned two important and surprising pieces of information: that our eyes see only a fraction of the light that reaches us from the Sun, and that nothing travels faster than electromagnetic radiation. They also learned that the kind of light used by garage door openers has a much lower frequency than the kind used by TV remotes, but Caroline's question remained unanswered — something was still missing, but she couldn't quite put her finger on it, so she decided to dig deeper on her own.

Caroline recalled the equation her father had shown her when explaining music—the one connecting wavelength (λ), frequency (ν), and speed—and thought that since light and sound are both waves, the equation should apply to light as well as to sound. The only difference would be the speed, denoted by c ("At last, an English letter!" she thought).

$$\lambda\nu = c$$

Even though the numerical values of wavelength and frequency for radio waves are different from those for infrared waves, their product is exactly the same: the speed of light, c. Caroline remembered that the wavelengths of audio, such as middle C, are a few feet, and she expected the wavelengths of infrared and radio waves to be much different. She wondered just how different they really are and decided to do a calculation.

Using SI units and the metric system, which she believed would make the arithmetic easier, she recalled the concept of scientific notation from her chemistry class: for example, a number like 10^8 means 1 followed by eight zeros, which is 100 million and 10^n where n is any integer means a 1 followed by n zeros. Since the speed of light is $3x10^8$ meters/sec, and a radio wave of 300 MHz, or 300 million hertz, can be written as $3x10^8$ Hz, the wavelength of a radio wave is

1 m/cycle =100 cm/cycle

$$\lambda = \frac{3x10^8 m/\sec}{3x10^8 cycles/\sec} =$$

"I didn't even need a calculator for that one. Between the metric system and scientific notation, it was a breeze—the distance from one peak to the next is 100 cm, or a little over 3 feet." Caroline is struck by the similarity between the wavelength of common audio, such as middle C, which has a wavelength of approximately 4 feet, and the wavelength of radio waves. She wonders if the similarity is just coincidence.

She calls her uncle who assures her that although her observation is interesting, the similarity is purely coincidental. He explains that the choice of radio wave frequencies is based entirely on practical considerations. Lower frequency radio waves can travel further and penetrate obstacles better and designing antennas with a characteristic dimension of a few feet is ideal for commercial use.

Satisfied with his explanation, she moves on to calculate the wavelength of near-infrared radiation, which turns out to be around 10^{-4} cm. Now she understands what Mr. Lewes meant when he said that the light waves used by garage door remotes are so different from those used by TV remotes. In fact, the ratio of their wavelengths is: 100/0.0001 = 1 million: This means radio waves are a million times longer than infrared waves.

The Time That Elapses Between Hearing and Seeing Visible Light Reveals the Distance of an Approaching Storm

Caroline begins searching the internet and before long confirms something Mr. Lewes had mentioned: all the energy that arrives from the Sun travels at the speed of light—186,000 miles per second. She still has trouble wrapping her mind around anything moving that fast. Sound, she remembers, travels at 1,126 feet per second at sea level. That's fast—more than ten times the speed of a car on the highway. But when she compares it to light, the difference is staggering.

She punches some numbers into her calculator.

"Light travels almost a million times faster than sound," she mutters, wide-eyed. "That's... mind-boggling."

Her uncle had explained that this wasn't just a curious fact—it had real-world consequences. Since light reaches us almost instantly, while sound lags behind, you can estimate the distance of a storm by counting the seconds between a lightning flash and the thunderclap that follows.

Caroline turns to Noreen. "So if someone hears thunder five seconds after seeing the lightning, how far away's the storm?"

Noreen thinks for a moment. "Well, if light gets here basically instantly, then it's the distance sound travels in five seconds, right?"

Caroline nods and taps on her calculator. "That's five times 767 divided by 3600... about 1.1 miles."

"Yeah," Noreen says, "my dad told me people used to use that all the time, before weather apps. Flash-to-bang, I think he called it. Pretty useful if you're caught outside—or even just near a window during a storm."

Caroline liked the phrase. "Flash to bang," she repeated. "Even if we don't need it anymore, it's still cool to know you can estimate something like that so easily."

Mary had been listening from across the room. "It's not just the number that matters," she said. "You'll see that same kind of thinking come up again and again in science—the idea that the timing or spacing of things can reveal something deeper."

But Caroline and Noreen were already off in their own world, scribbling numbers and speculating how long it would take sound to reach them from different distances.

Caroline grinned. "OK, now I kind of *want* a thunderstorm, just so we can try it."

Noreen laughed, then paused. "You know, I still don't really get what makes thunder. I mean, I know it's connected to lightning—but how does lightning *make* sound?"

That made Caroline pause too. She looked toward her mom. "Yeah... how *does* that work?"

Mary smiled. "Good question. Let's talk about it, I think you'll both be surprised by the answer.

The Air Surrounding Lightning s Hotter Than the Surface of the Sun

"To begin, let me ask *you* a question: what do you suppose the temperature is of air surrounding a lightning bolt?"

Caroline and Noreen guess that the air is perhaps 100°C, which isn't unreasonable, given their limited knowledge of the temperatures of objects in their surroundings. At this point, they don't even know how hot a lightbulb filament is, but if they had known more physics, they might have been able to make a better guess based on the color of the glowing filament.

Lightning Creates Shock Waves

Mary: "I think most people would guess something similar, but it seriously underestimates the temperature of the air surrounding a lightning bolt. In reality, *lightning rapidly heats the air to temperatures as high as 28,000°C.* To put that in perspective, it's much hotter than the Sun's surface temperature, though cooler than the Sun's interior. This extreme heat causes the air to expand very rapidly, just as it does in an explosion. And just as in an explosion, the rapid change in pressure creates a shockwave—similar to the sonic boom heard when a plane breaks the sound barrier—that travels outward as sound waves, which we hear as thunder."

Later that evening, as her parents are about to catch up on the news of the day, Caroline turns to her mother and says, in a slightly frustrated, childish tone—after all, she only recently turned 15: "You're just like Dad, always using words I don't understand. Now I have another question: what's a sonic boom? And please don't tell me it's a loud noise a plane makes when it breaks the sound barrier—I already know that."

Mary takes a deep breath and patiently reminds her daughter that the pressure waves—the sound waves—created by an aircraft moving at less than the speed of sound, spread out in all directions, but are more compressed in front of the plane than they are toward the rear. Then realizing that she'll need to keep the explanation simple she asks her daughter to imagine a point source of sound rather than a large complex aircraft.

Mary: "Initially the source is moving at less than the speed of sound and moving toward an observer. The source is also accelerating, and as it does the distance between successive pressure waves in the direction of motion decreases. For both reasons the frequency of sound perceived by an observer increases. But let's just focus on the increase due to acceleration.

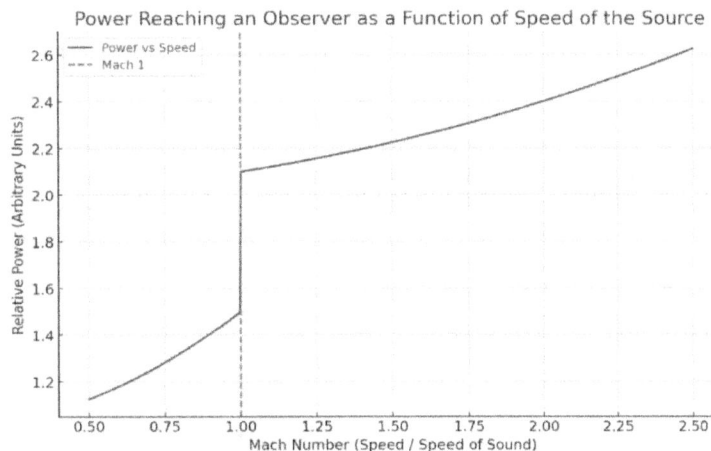

Fig 1.Results of a computer simulation showing sonic power reaching an observer as a function of the speed of the source, starting with subsonic velocities (less than Mach 1), passing through Mach 1, and continuing to accelerate. In the subsonic region the power gradually increases as speed increases. At Mach 1 there is a sharp increase in power due to the formation of the shock wave (the sonic boom), representing a spike in energy reaching the observer. In the supersonic region (above Mach 1) the power continues to increase but at a slower rate compared to the sharp jump at Mach 1.

"Since each sound wave carries energy, and the number of wavefronts reaching the observer per unit time also increases as the source accelerates, the energy per unit time (i.e., sound intensity) reaching an observer increases. As the source continues to accelerate the distance between successive wavefronts at its front decreases, while the observed intensity level increases. *As the source approaches the speed of sound (Mach 1), the wavefronts that were emitted at different times begin to coalesce. At Mach 1 they're essentially stacked on top of one another and constructively interfere so that multiple sound waves emitted at different times reach an observer simultaneously. The result is a discontinuity in pressure, temperature and density—what we call a shockwave.* This shockwave is perceived as a sudden, intense thump-like sound generated by the rapid change in pressure."

At that point Mary shows the girls a computer simulation she had generated as an undergraduate student, showing that the power reaching an observer as a function of speed changes discontinuously at Mach 1.

The Frequency of an Electromagnetic Wave Has a Profound Effect on Its Behaviour

Mary emphasizes that to keep the explanation as simple as possible she has omitted many details. Caroline seems to have grasped the main ideas and feels that she's learned a lot from this brief lesson, but she continues to struggle with putting the speed of light into perspective and connecting it with things she understands. Then she recalls that the circumference of the Earth at the equator is approximately 24,000 miles and quickly calculates that light could circle the Earth nearly eight times in one second.

"Incredibly fast!" she thinks. Her mind floods with thoughts: "I don't know why I picked that example—light doesn't bend," and almost simultaneously another thought strikes her: "Sammy, the boy up the street, is a ham radio operator and hears broadcasts from all over the world. Do radio waves bend? Or are they somehow reflected? They're the same as light except for wavelength—I'm confused. I'll ask Dr. Kahn, my science teacher. I'm sure he'll know the answer."

Caroline told her mother that Dr, Kahn did provide some useful information. He explained that shortwave radio waves can circle the globe because they are reflected by the ionosphere, an upper atmospheric layer of electrons and ions created when extreme ultraviolet radiation from the Sun knocks electrons from nitrogen, oxygen, and other molecules. Reflection only occurs for electromagnetic radiation with frequencies below a critical value known as the plasma frequency, and radio waves meet this condition. The *reflected waves then bounce between the Earth's surface and the ionosphere, a phenomenon called skywave propagation, which enables long-distance communication.*

Caroline: "When I asked Dr. Kahn why higher frequency radiation such as light can't circle the globe the way radio waves do, he said that electrons can't respond quickly enough to follow the oscillations of radiation whose frequency exceeds the plasma frequency—they're slowed by inertia and continual collisions with ions and neutral atoms. Since the frequency of visible light exceeds the plasma threshold, it passes unimpeded through the ionosphere unless it's absorbed by nitrogen dioxide or scattered by aerosols in the lower atmosphere.

"He said he didn't expect me to understand all the details but hoped I'd get the general picture. And he was right—I don't understand why a collection of ions and electrons at a certain density and mass can reflect electromagnetic radiation, or why the plasma frequency exists, especially with its sharp cutoff. He said that a few more years of study would probably be needed to fully grasp such concepts, but I'm grateful he provided an explanation that was clear enough to start building a deeper understanding."

Then suddenly, as she was recounting her conversation with Dr. Kahn, she had a new thought that was strange to her—it was more philosophical than scientific: "sometimes just knowing is enough, and sometimes my state of knowledge will force me to accept that knowing is enough". And then in typical fashion she concluded, half in jest "at least until tomorrow."

So far, Caroline's search for an answer to why infrared remotes must be pointed at the receiver, while garage door remotes do not, has been both frustrating and exhilarating. It's exhilarating because she's beginning to see a world that is more nuanced and complex than she had ever imagined. She's learned that we are all essentially blind to light, detecting only an extremely small fraction of the radiation constantly moving around us; that light travels a million times faster than sound; and that even though light and sound are both waves, they are different kinds of energy. Yet it's frustrating because the initial question remains unanswered. It must have something to do with their frequencies, but what specifically about the frequency difference might explain it?

This is leading me nowhere. I'm going to have to ask Mom or Dad. Dad's rehearsing today, so I'll ask Mom—besides, it should be OK to interrupt her; she's only thinking about money.

Caroline told her mother about her discussion with Mr. Lewes and how she and Noreen were trying to understand why TV remotes have to be pointed, but garage door remotes do not. She said she learned a lot of interesting facts but still didn't know the answer to her question.

Mary unfortunately couldn't help. She reminded Caroline that nearly two decades had passed since she graduated from college, but she promised to give the question some thought and to discuss it again in a few days.

Several weeks later, Caroline finally sits down with her mother.

Caroline: "Wow, Mom, that's some two days—it's been closer to a month."

Mary: "Sorry, Caroline. Things have been hectic at my office, and there was a lot of money at stake."

Caroline: "Mom, did you ever hear Money, Money, Money by ABBA? Someday I'm going to sing it to you."

Mary chuckles: "Caroline, someday you and I are going to have a long talk about finance and economics. It's true that without science and technology we wouldn't have a modern society, but it's also true that modern society wouldn't be possible without money. It's one of the most important human inventions, right up there with writing. More immediately, we wouldn't be nearly as comfortable as we are if my salary weren't as high as it is."

Mary senses Caroline is reflecting on her words, but her daughter remains silent. She smiles at Caroline and begins by reviewing what they know. *Caroline observed that TV remotes have to be pointed, while garage door remotes do not. The difference lies in the electromagnetic frequencies they use. The wavelength of radio waves used by a garage door remote is in the range of approximately 3-6 feet, while the wavelength of infrared light used by TV remotes is approximately a million times smaller.* However, Caroline's question still remains.

Mary attempts a deeper answer. She explains that the reason garage door remotes don't have to be pointed is related to the ability of radio waves to penetrate objects that are opaque to infrared light, such as bricks, concrete and other hard materials (if they're not too thick). Infrared light, like visible light, can't even penetrate the plastic casing of the remote. The remote emits light only through a small opening that is transparent to infrared, so it must be pointed directly at the TV receiver.

Caroline feels that she's finally starting to understand. Nevertheless, she responds, almost as if in an involuntary reflex, with the question that strikes fear into her parents' hearts: "I'm sorry," she says sweetly and somewhat apologetically, "I don't understand."

Mary: "Yes, sweetheart, tell me."

Caroline: "If they're both electromagnetic radiation and travel at light speed, why are radio waves unaffected by objects that can't be penetrated by infrared radiation?"

Mary then goes a little deeper saying that molecules of which the plastic is made absorb electromagnetic energy having frequencies in the infrared range, but don't absorb long wavelength radiation like radio waves.

She summarizes saying that **wavelength determines how waves interact with obstacles**. Radio waves (wavelengths in the meter range) diffract around obstacles and pass through materials that do not strongly absorb them while infrared waves (micrometer-scale wavelengths) are much more likely to be absorbed by molecular vibrations in materials.

An Electric Current Produces a Magnetic Field

Caroline: "My teacher spoke briefly about electromagnetic waves, and you, Dad and Mr. Lewes have all told me a lot of facts, but I still feel that I'm missing something important."

Her mother responds by explaining that electromagnetic waves, their motion, and the mathematics needed to describe them involve deep concepts that neither she nor Don know much about.

Don then suggests that Caroline speak with her Uncle.

Caroline: "That's OK with me; when do you think I can speak with him?"

Don: "He tends to rise early, around 4:30 or 5, to have some uninterrupted time to develop and explore the mathematical consequences of his experiments and to plan his day. So, around 7 AM our time, for half an hour or so before you leave for school, might work. Why not text him to arrange a Zoom conversation?"

A few days later Caroline is sitting in front of her computer talking with her uncle.

Zack: "Where to begin? What I'm going to say will lack detail, but it should give you an understanding of the relevant concepts. A good place to start is with an observation made in the early 19th century by the Danish physicist Hans Christian Ørsted. He noticed that *a compass needle is deflected in the presence of a conducting wire carrying a constant (non-alternating) electric current,* suggesting a fundamental connection between electricity and magnetism. Shortly afterward, Biot and Savart developed a mathematical theory relating electric current to the strength and magnitude of the magnetic field as a function of distance from the wire.

"Now, here's a question for you to think about—one that might come as a bit of a shock: Suppose you were sitting on top of an electron in that wire, that you were in a frame of reference moving with the current and could measure the strengths of the electric and magnetic fields? What do you think your instruments would tell you?"

Caroline: "I don't know—but I'm pretty sure I'd see the electric field."

Zack: "What else?"

Caroline: "Well, I'm moving with the current, and as far as I'm concerned, there's no current flowing by me. So, if a magnetic field arises because of a current and I can't detect a current, I guess I wouldn't detect a magnetic field. That's weird—can it be that the extent to which I observe a magnetic field depends on my speed relative to that of the current?"

Zack: "Exactly. You've just discovered a fundamental result from special relativity. *When you talk about detecting an electric current, it means that as a stationary observer you see a flow of electrons that generates a magnetic field. However, if you're moving with that flow, the electrons appear to be stationary, there is no current relative to you, and you will not detect a magnetic field—only the electric field.* If your speed changes relative to the speed of the electrons, the field that was purely electric in the electron's frame of reference transforms into a combination of electric and magnetic fields in your new frame of reference. The energy that was initially all in the electric field is now distributed between the electric and magnetic fields, and the extent of this redistribution depends on the relative velocity between the observer and the charge carriers (electrons)."

Caroline: "What do you mean by the energy of an electric field?"

Zack: "The concept of electric field energy is also quite deep, but some intuition can be gained by considering gravitational field energy. I know you've heard of the potential energy of an object in a gravitational field. For example, if you hold a ball or any object at some height above the ground, it has potential energy—it's the energy the object will acquire when it falls to the ground. You could say the gravitational field has energy in the sense that it can impart kinetic energy to a mass. An electric field works similarly: it can impart kinetic energy to a charge."

Maxwell and Faraday Developed the Principles of Electricity and Magnetism that Transformed Society

Caroline: "OK, I think I understand."

Zack: "A decade later, a brilliant scientist named Michael Faraday—the self-educated son of a blacksmith—made one of the most consequential discoveries of the 19th century. He noticed that *when a magnet is moved in and out of a loop of conducting wire, a current is induced in the wire.* And, as Ørsted demonstrated, a magnetic field surrounds a current-carrying wire. Are we following so far?"

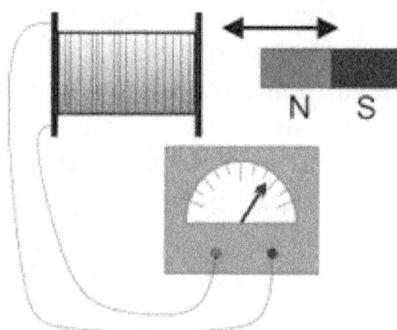

Figure 2. When the bar magnetic is moved in and out of the conducting coil, a current is induced and registered by the galvanometer.

Caroline: "Yes, so far so good."

Zack: "Now, here's the real breakthrough: when the movement is oscillatory—that is, back and forth in a periodic motion—the current will also be periodic, and consequently, the magnetic field will...?"

Caroline: "Oscillate."

Zack: "Yes, exactly. And if another closed-loop conductor is placed nearby, an alternating current will be induced in that loop. The magnitude of that current, according to Biot and Savart's theory, will depend on several factors, but perhaps most importantly, on the orientation of the two loops. It's maximum when they're perpendicular and zero when they're parallel. And please don't ask me to explain the angular dependence right now; it would take us too far afield, and you should be able to research the answer yourself."

Zack then continues, "Speaking of going far afield, Faraday's observations led to the concept of electric and magnetic fields filling the space around the wires. It's these fields that a charge responds to, rather than responding to a force acting at a distance."

"Uncle Zack," Caroline says teasingly, "you have the same corny sense of humor as your brother."

Her uncle smiles but continues without acknowledging her comment: "This idea of interrelated, oscillating electric and magnetic fields provided the empirical basis for Maxwell's mathematical theory, which unified electricity and magnetism. Of course, I'm leaving out a lot—the concepts are highly mathematical—but I expect you'll be ready to take a deep dive in a few years. One last thing, and it's of very practical significance: *Maxwell's unified theory of electromagnetism, combined with Faraday's experiments and the concepts he introduced, laid the groundwork for converting kinetic energy—such as from a wind turbine or a waterfall—into alternating current.* To say this was consequential would be an understatement of the highest order."

Caroline: "I guess..." she says, with a slight inflection, "but I don't understand. How is kinetic energy converted into electrical energy?"

Zack: "When a magnet is moved back and forth through a loop of conducting wire, the changing magnetic field induces an alternating current (AC) in the wire. However, Faraday's laboratory demonstration, where he provided the mechanical energy to move the magnet, doesn't address how electricity can be generated on a scale sufficient to power the planet. Where would the mechanical energy come from?"

Caroline: "Windmills and waterfalls."

Zack: "Yes, those are two of many possibilities. The idea is to harness the massive amounts of kinetic energy found in falling water, wind, or even steam (which can be generated by a nuclear reactor). We won't have time to go into the details of how that's done right now, but the general principle is common to all three possibilities and is relatively simple.

"Let's take wind turbines, or windmills as they're often called, as an example. For now, let's set aside the question of how much power a typical turbine can generate on a typical windy day. It's not difficult to estimate, and we'll get to that in another meeting. In the meantime, feel free to try it on your own.

"The process begins with the conversion of wind's kinetic energy into the rotational kinetic energy of the large windmill blades. The energy of the rotating blades is transmitted to a shaft through a gearbox that increases the shaft's rotational speed, which is necessary because the blades rotate relatively slowly compared to the speed needed by the generator. The shaft drives a generator, where a rotating magnet induces an alternating current (AC) in stationary coils of wire. The resulting AC is then converted to a specified voltage, which is transmitted via the electrical grid, delivering power over large distances. This description is somewhat perfunctory, but I think you get the picture, and I'm sure you can fill in the gaps with a little research.

"Incidentally, do you see the difference between how wind turbines generate electricity and how waterfalls do it?"

Caroline: "Yes, I think I do. Wind turbines use the kinetic energy of the wind to drive the blades of a turbine, whereas hydroelectric power uses a reservoir of water that falls in a controlled manner, converting gravitational potential energy into the kinetic energy that drives the blades of a turbine."

Zack: "That's great—you've got the picture. Let's follow up in a couple of weeks with some quantitative comparisons between different energy sources."

Caroline: "Thanks, Uncle Zack. I really appreciate the clarifications."

That evening, after dinner, Don and Mary asked Caroline about her conversation with Zack.

Caroline: "It was great; I really learned a lot."

Mary: "I remember being astounded when I first learned that *electromagnetic waves are all around us*. Visible light is an electromagnetic wave, so is infrared (heat) and ultraviolet light, and so are radio waves and microwaves. They're all electromagnetic waves. They all travel at the same speed and differ only in their frequency. *They crisscross every which way, and we're completely oblivious to their presence, except for a very thin sliver of the spectrum: the visible portion, which receptors in our eyes respond to, and the infrared portion, which receptors in our skin respond to.*"

Don: "*Similar to sound waves, electromagnetic waves transfer energy in a forward direction; however, unlike the oscillating air molecules in a sound wave, the oscillating electric and magnetic fields of an electromagnetic wave do not move forward with the wave, but oscillate perpendicular to its direction, and unlike sound, electromagnetic waves do not require a medium to propagate.*"

Caroline: "What??"

Mary: "Yes, I know it's surprising but an explanation would take a while so it's best to postpone until the weekend. Besides, you've had a full day and you still have homework to complete."

Caroline, with a disappointed tone, indicates with no ambiguity that she finds homework boring compared to the conversation they're having. Nevertheless, she manages to discipline herself and says a resigned good night as she leaves the living room.

It's now Saturday morning. The Angstroms have no sooner sat down to breakfast when Caroline reminds her parents about the conversation that they were having earlier in the week about electromagnetic wave propagation.

Her father responds: "Think about the light and heat from the Sun that you see and feel almost every day—they travel through empty space, don't they?"

Caroline: "That seems weird."

Don: "I know, and yes, for many years, people believed there was no such thing as empty space. *The planets, the Sun, and all the matter in the known Universe were thought to move through an invisible substance called 'ether,' a medium that enabled electromagnetic waves to propagate, just as air enables the propagation of sound.* If you have the time, you'll gain real insight into optics by reading about how the nonexistence of the ether was proved by a brilliant experiment.

"Here's another little tidbit for you to investigate: the oscillations of the magnetic and electric fields are perpendicular to one another and to the direction of motion of the energy wave. In other words, they're transverse to the direction of motion. Sound waves, as you'll recall, are longitudinal—they vibrate in the direction of motion."

Caroline: "I guess there's a lot of commonplace behavior that I have no clue about."

Microwaves Cause Water and Other Polar Molecules in Food to Vibrate and Rotate, Producing Heat Through Friction and Molecular Motion

Don: "We'll get back to these concepts in more detail, but for now, you can probably recognize that the intellectual and practical importance of *Maxwell's equations* is staggering. They *form the foundation for wireless communication, laser design, medical imaging, even cooking—essentially every technology that uses electricity and magnetism.* Maxwell died before the age of 50, but you can count on one hand the number of physicists who have had his impact on the world."

Caroline: "Cooking?"

Mary: "Yes, the microwave oven. It's one of many examples of commonplace technologies, based on electromagnetic radiation, that we take for granted. Did you ever wonder how it heats food?"

Caroline: "Yes, you place the food in a microwave oven and press a button."

Mary half smiles, and then says, "Yes, that's what happens, but it doesn't explain how it happens."

Caroline: "I know, Mom—so how does it happen?"

Mary: "The microwave energy is transferred very efficiently to food, and the reason it's so efficient is that food is mostly water, and water absorbs microwave radiation."

Caroline: "Why is that?"

Mary: "Why did I know you were going to ask why?"

Caroline: "I don't know, why?"

Mary: "Do you remember learning in science class a few months ago that a water molecule consists of two hydrogen atoms connected to an oxygen atom? Well, the hydrogen atoms have a slightly positive charge, and the oxygen atom has a slightly negative charge—so you can think of the water molecule as a dumbbell: positive charge at one end and negative charge at the other end, what scientists call a *dipole*. The dipole can vibrate and bend and rotate, but with respect to the transfer of microwave energy it's the rotational motion that matters. *The rapidly oscillating microwave field interacts with the dipole, forcing the water molecules to continuously attempt to realign. This motion leads to intermolecular friction and heating, which warms the food.*"

Caroline: "OK, I understand that, but I still don't understand why bonds vibrate, or how the particular atom types that they connect determine their frequency."

Mary thinks to herself, "I guess another why is always possible," feeling a bit exasperated.

Caroline has an approximate picture of how microwaves heat food but decides not to attempt a deeper dive until she learns more physics. For the same reason, she decides to let go for now of those other questions—why bonds vibrate and why they vibrate at particular frequencies. She feels satisfied that she has explored at least three levels of "why." But why does she feel satisfied when she usually doesn't? She thinks for a moment and realizes that although she has more questions, the simple question that started this journey of inquiry led her in unexpected directions and toward an understanding she never could have imagined. Her view of the world has been forever changed, and she experiences a sense of satisfaction she has never felt before.

Conversation 6

Seeing Through Objects

Caroline's English teacher gives a fascinating lecture on Superman as a Romantic hero, but Caroline is more intrigued by Superman's miraculous abilities than by Romanticism. Among the many questions that come to her mind is how Superman's X-ray vision works and why he can see through everything except lead. Curious, Caroline discusses these questions with her uncle and teachers and supplements their insights with her own research. Her exploration leads her to uncover not only the answer to her question but also to the related question of why glass is transparent to visible light while wood, metal, and many other common solids are not. In the process she learns about the photoelectric effect which converts light into electricity, and Einstein's revolutionary conclusion that light behaves as particles called photons. Einstein's startling announcement illustrates the conservative nature of science and the difficulty even brilliant minds like Max Planck face in letting go of existing paradigms, such as the belief that light is strictly wave-like. Caroline struggles to understand how light can behave as both a wave and a particle. She is stunned by the implications of wave-particle duality, including the many-worlds theory, which strikes her as something out of science fiction. Despite the bizarre nature of light, Caroline begins to appreciate its power as a tool for scientific discovery. She learns how it can be used to probe the structures of DNA and proteins, which provide the foundation for modern biotechnology. Her journey leaves her with a sense of wonder about the mysteries of the Universe and the transformative role of science in uncovering them.

Superman Would Have Been More Realistic with Infrared Vision

One day early in her second year of high school, Caroline arrived home with a newfound enthusiasm—her English teacher, Dr. Siraganian, a genuine literary scholar and one of the few faculty members in her high school with a Ph.D., had taught a lesson about one of the Superman movies.

Caroline: "I never thought much about comic books, and I assumed they were read only by people who had no interest in serious literature."

Mary: "I'd say your assumption is mostly correct, but can we wait until your father comes home before we discuss what you learned—I'd like him to hear what Dr. Siraganian said that has made you so enthusiastic."

Soon after, the family sat down for dinner, and Caroline launched into the discussion without hesitation.

Caroline: "Dr. Siraganian began his lecture by distinguishing between Romanticism with a capital 'R'—which embodies the artistic movement focused on emotion, individuality, and the sublime—and romance with a small 'r,' which is what most of the kids in my class fantasize about..."

Don (interjecting with a cautious smile): "You needn't say more. I know exactly what you mean."

Caroline: "He really didn't go so far as to portray Superman as the embodiment of Romanticism, but he did say that Superman symbolized some of what Romanticism was about. And of course, there were hints of an unfulfilled romance between Superman and Lois Lane, but he said those scenes were mainly there to add some spice to the movie."

Mary: "That's very interesting, but can you say something more concrete?"

Caroline: "Well, I really haven't been able to assimilate everything he said, but I did take lots of notes—so here goes.

"According to Dr. Siraganian, the Romantic hero is a complex figure, torn by inner struggles and intense emotions. He pointed out that in the 2006 movie *Superman Returns*, Superman returns to Earth after a five-year absence during which he traveled to what remained of his home planet, Krypton, after its tragic self-induced destruction. Those five years were a period of self-discovery during which Superman grappled with his identity, his purpose, his responsibilities, and the burden of his powers. In the end, he committed with all his being to serving humanity. Dr. Siraganian said that Superman's journey reflects the Romantic theme of individual struggle and personal sacrifice for the greater good.

"Superman's personal qualities and his struggle to find his purpose are not unlike the struggles of Romantic heroes, including Achilles, Hamlet, Othello, Macbeth, the knights in *The Faerie Queene*, and many others. These figures often display the characteristics of Romantic heroes: courage, passion, and a sense of moral integrity. They grapple with inner conflicts and embody the Romantic notion of the individual fighting against powerful forces, both external and internal. This aligns well with the Romantic emphasis on heroism and the championing of noble causes.

"Dr. Siraganian also mentioned that a focus on an individual's experiences and inner feelings, as well as an emphasis on personal journeys and self-expression, is a central feature of Romantic literature. He explained that Romanticism incorporates elements of the mysterious, the supernatural, and the extraordinary. Naturally, Superman, as a supernatural being with extraordinary powers, fits this mold. He represents an idealized version of humanity, symbolizing the pinnacle of human virtues: strength, kindness, and moral integrity. This idealization is a common feature of Romantic literature. All in all, Dr. Siraganian argued, Superman embodies the Romantic hero—an exceptional figure with a strong moral compass who fights for the greater good."

Caroline reflected on these ideas, wishing she had a better understanding of literature to fully grasp Dr. Siraganian's allusions to the heroes of both the Romantic and pre-Romantic periods. However, she took comfort in his reassurance that understanding every detail wasn't necessary at this stage. His goal was to provide a broad perspective—a ten-thousand-foot view—of the connections between the Romantic hero and Superman's life and character. He promised that later in the semester, the class would dive deeper into Romantic philosophy through the works of some of the great writers of the period.

Don smiled appreciatively. "He sounds like a great teacher—someone who stimulates your imagination to see what you haven't seen before and leaves you wanting to learn more."

Caroline nodded in agreement, reinforcing her father's words. She admitted that she had never realized how literature captures universal truths that transcend geography and history.

"People alive today and those who lived hundreds of years ago share a common psychology, from which our best and worst instincts spring: love and hate, generosity and greed, peace and war, jealousy and admiration." She paused thoughtfully before shifting the conversation. "Although I feel elated by this new way of thinking, I do have a question," she said, with a small laugh. "It's not directly related to Romantic heroes, but when I think of Superman, I think of his powers—especially his X-ray vision."

Mary chuckled. "It would be unlike you not to have a question. Go ahead and ask; maybe we can steer you toward an answer."

Figure 1. X-rays appear to emanate from Superman's eyes. Caroline says this is extremely implausible because there are no known mechanisms by which an organic visual system could emit x-rays, nor are there any known instances.

Caroline had intended to ask how Superman's eyes could produce X-rays, but as she spoke, she realized that Superman's eyes don't produce X-rays, they detect them. Unlike ordinary human eyes, which detect visible light, Superman's eyes can sense X-rays, a higher frequency part of the electromagnetic spectrum.

Caroline: I guess the reason Superman can see through walls is that X-rays emitted by objects on the other side of the wall penetrate it, just as they penetrate soft tissue. But how can he see a person or any other object on the other side of the wall unless those objects emit X-rays. I know we emit heat, which is also electromagnetic radiation, but do we, or any products of the human imagination except for the devices designed by engineers, emit X-rays?"

Neither Mary nor Don knew of any evidence that living organisms or other common objects emit X-rays.

Caroline's response, however, indicated that despite the conversation they just had she still didn't quite grasp the ideas presented by her teacher:

Caroline: "If X-rays are emitted by neither the living nor the non-living, then eyes capable of detecting X-rays wouldn't help Superman in the least—and that's why I struggle with comic books. The science behind them often doesn't make sense. It's difficult for me to be enthusiastic about internally inconsistent stories."

Mary: "Caroline, you're taking the Superman saga much too literally. We just completed a fairly detailed discussion of Dr. Siraganian's lecture and concluded that the Romantic superhero is a symbol. Although, I'll admit, the saga would have been more sensible if Superman's creators, Jerry Siegel and Joe Shuster, had given him infrared rather than X-ray vision."

Don then added that the great English poet Samuel Taylor Coleridge suggested that a reader might willingly ignore the implausibility of a narrative if not ignoring it would prevent the reader from feeling the full impact of a powerful truth—a practice he referred to as a willing suspension of disbelief.

Caroline frowned, mulling over her father's words, and finally asked whether the suspension of disbelief could apply even to an illogical narrative. After receiving an affirmative answer, she decided to let the issue drop and move on, allowing for the possibility that she might need some help analyzing literature.

Even as Caroline spoke those words, a flood of thoughts swirled through her mind. The questions came quickly, and then some of the answers seemed to come from nowhere, and then— frustratingly—came more questions. Why, she wondered, can't humans see through walls, or through wood, or through brick, or through so many other forms of matter? And again, almost as if from nowhere, the answer flashed through her mind: humans can't see through walls or wood because these materials absorb visible light. If an object doesn't emit or reflect light in the visible spectrum, it's invisible to our eyes. However, this realization led to another question: Why doesn't glass absorb visible light?

Mary, who was of course oblivious to what Caroline was thinking, continued the previous conversational thread. She explained to her daughter that people who move civilization forward, whether in science or in the arts, do so with great effort—but recognizing and understanding the achievements that demonstrate the artistic and intellectual heights humans can reach also requires great effort.

Mary: "So be prepared to work at it, be patient, enjoy learning, and I'm sure your appreciation will grow."

Roentgen Discovered X-rays and Demonstrated Their Potential for Enabling a Medical Revolution

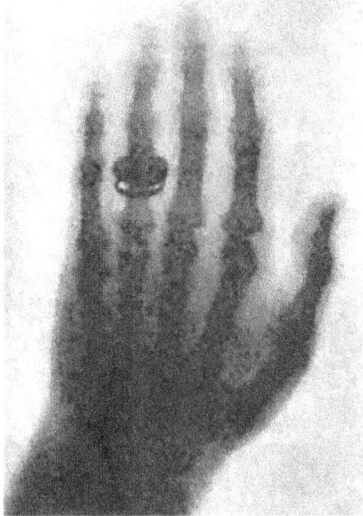

Figure 2. One of Roentgen's photographic plates showing his wife's hand and wedding ring. The reversal of black and white in the image is the result of chemical processing. Modern methods would show the blocked regions in white, and the unblocked in black.

Caroline then said what she was thinking, although it appeared to her parents that she was changing the conversation.

"Last year we learned that X-rays are a form of electromagnetic radiation – they're waves just like visible light except that their wavelength is roughly 100,000 times shorter, which is much too short for our eyes to detect. Our teacher said that when Roentgen, a physicist who lived more than 100 years ago, discovered this new form of radiation, he suspected that it could be used to take photographs, just as visible light is used to take photographs. To test the idea his wife placed her hand between an X-ray source and a photographic film, and when the film was developed it was white where the X-rays were blocked by bone, gray where they were partially absorbed by soft tissue, and black where they passed through air unimpeded."

As Caroline spoke, she again realized that she didn't fully understand why visible light can pass through some substances and not others. Now, she was puzzled by the results of Roentgen's experiment: X-rays penetrate soft tissue more easily than they pass through bone. Of course, the adjectives "soft" and "hard" seemed to hint at the answer, but they were only words. Perhaps it really is just a matter of different densities—or could it be that X-rays interact differently with the molecules in soft tissue than they do with the molecules in bone?

Don, who was surprised at how difficult it was to answer Caroline's question, replied in a way that was becoming increasingly common: "Try asking your uncle; I'm sure he can be helpful."

Mary: "I'm also curious—so while you're at it, can you ask Uncle Zack or Aunt Zoe how soft tissue absorbs X-rays and how it differs from absorption by bone?" In fact now that your Aunt has moved to New York, it will be easy for us to get together more often.

Caroline: "It must feel good to have your sister nearby. What made her move."

Mary: "As you know your aunt is single, and there isn't much of a social life for singles in Lewiston, so even though she thought Bates was an excellent college, when The City College of New York offered her a faculty position, she jumped at the opportunity. And yes, we both feel good about being nearby; we were always very close.

Caroline: "Well I look forward to talking with her and with Uncle Zack. I'll let you know what they say."

It was a Saturday morning, a few weeks later, when Caroline and her parents sat down for breakfast. Caroline shared what she had learned from her aunt and uncle about various topics related to the absorption of electromagnetic waves. Her Uncle Zack had suggested that, for the purpose of understanding absorption, it would be easier to *think of light as particles, which scientists call photons, that have different frequencies for different colors.* When Caroline expressed confusion over how a particle could have a frequency, Zack explained that even scientists didn't truly understand it, and that light sometimes behaves as a wave and at other times as a particle, depending on the nature of the experiment.

Caroline: "Uncle Zack reminded me of something I learned in science class last year: that it's sometimes helpful to *picture an atom as a miniature solar system.* At its center stands a nucleus composed of neutrons and protons, with electrons orbiting about it much as planets orbit the Sun. Unlike planets, however, *electrons are restricted to specific orbits, meaning only a limited number of energy levels are available to a bound electron.* This means that when an X-ray, which typically has an energy 100 to 1,000 times greater than visible light, collides with bone, it can knock a tightly bound calcium electron out of its orbit, with the electron absorbing only a fixed amount of the X-ray's energy."

Einstein's Paradigm-Shifting Theory of the Photoelectric Effect Laid the Foundation for New Industries

Caroline continued, "Uncle Zack then shifted the topic slightly, explaining that the *interaction between atoms and the photons that dislodge electrons from their orbits is known as the photoelectric effect, a term that is most commonly used when visible or UV light releases electrons from metal surfaces* rather than from bone. He went on to say that the phenomenon was first observed by Heinrich Hertz, but it raised puzzling questions. For example, *electrons are ejected only if the frequency exceeds a metal-specific threshold,* but the wave theory of light predicts a gradual increase in the number of emitted electrons as frequency increases. In addition, *the number of electrons ejected is independent of brightness*, which also can't be explained by the wave theory.

"Uncle Zack said that these findings baffled scientists for decades until Albert Einstein, who was a 26-year-old patent clerk, proposed a revolutionary mathematical theory based on the idea that optical energy is carried by particles, now known as photons, rather than waves, and that the energy of a photon is proportional to frequency."

Don: "Yes, as I recall Einstein's theory followed *Max Planck's earlier theory, which stated that electromagnetic energy is transmitted in discrete packets, or quanta, and that their energy is an integer multiple of their frequency, v.* In other words, energy could be hv, 2hv, 3hv, and so on, where h is a proportionality constant named in honor of Planck. However, energy could not be a non-integer multiple of hv, like 1.2hv."

Caroline: "Uncle Zack said that Einstein was awarded the Nobel Prize for explaining the photoelectric effect—it marked a monumental shift in our understanding of light, challenging centuries-old scientific paradigms."

Don: "Scientific paradigm? Wow, that term itself conveys a profound concept. Do you know what it means?"

Caroline: "Uncle Zack sometimes reminds me of you—he uses words I don't always understand. But when I asked him about it, he told me about a historian of science named Thomas Kuhn, who introduced the term in his book *The Structure of Scientific Revolutions*. Kuhn explained that science typically advances slowly and steadily, but occasionally there is a dramatic shift in how some aspect of the world is perceived—like the realization that the Sun, rather than the Earth, is at the center of the solar system, or that humans evolved from other species instead of being created as they are, by God. Kuhn called these shifts departures from an existing paradigm. Einstein's theory, which states that light exchanges energy with a metal in discrete units rather than continuously, was an example of such a shift.

"Uncle Zack also mentioned that the photoelectric effect forms the basis for solar panels, which convert sunlight into electricity and are a major technology in the quest for alternatives to fossil fuels."

Don: "Yes, *the photoelectric effect, the conversion of light into electricity; underpins many everyday technologies, including photocells, digital cameras, and barcodes—it's virtually everywhere.*"

Caroline: "I'm really surprised that even though I and just about all of my classmates know Einstein's famous equation $E = mc^2$, which expresses the equivalence of mass and energy, I never knew about the photoelectric effect and his explanation—for which he was awarded the Nobel Prize."

Don: "Yes, people are often surprised that Einstein was awarded the Nobel for his explanation of the photoelectric effect, a fundamental concept in physics, rather than for relativity. While the photoelectric effect is a standard topic in physics, the details of Einstein's explanation, which clarified the quantum nature of light, are less widely understood by the general public.

"Einstein first became well known to the general public when a striking prediction of General Relativity—the bending of light in the presence of a massive object—was verified. He predicted not only that it would bend, but precisely how much it would bend when passing by the Sun. This prediction was supported by observations made during a 1919 solar eclipse by a team led by Sir Arthur Eddington. These observations are part of a broader body of evidence that has since confirmed the bending of light. The news of this confirmation catapulted Einstein to international fame, making him a household name synonymous with genius.

"$E = mc^2$, derived from his Special Theory of Relativity, predates General Relativity by over a decade. While it wasn't immediately well-known to the general public, its significance became apparent with the advent of nuclear energy. The equation's fame isn't solely attributable to nuclear weapons, though their development certainly played a role. The equation's elegance and its revelation of the fundamental relationship between mass and energy also contributed to its prominence."

Caroline: "My English teacher, Dr. Siraganian, said the equation underlies, or perhaps symbolizes, a profound change in the way humanity sees itself. He said that the twentieth century was split in two: before and after nuclear energy, or to put it more starkly, before and after nuclear weapons; before and after the recognition that the human race has the power to destroy itself. That knowledge, that bite of the apple, as Dr. Siraganian put it, uncovered a fracture in human morality, with the counter currents between the moral precepts that most humans subscribe to, and the amoral tendencies driven by geopolitics, on full display. Unfortunately, the latter holds sway, leaving the human race in a position that can be described euphemistically as very fragile."

Caroline continued: "Dr. Khan mentioned that Oppenheimer and other physicists recognized the danger posed by nuclear weapons and made proposals to try to prevent their proliferation, but they weren't accepted."

Don: "The story of nuclear proliferation illustrates a general truth, almost a law, about technology: once a useful technology has been developed, it will be widely adopted and, in the process, made more powerful. It's a truth recognized even in ancient Greece, encapsulated in a famous tale you're surely familiar with."

Caroline: "You mean Pandora's Box?"

Don: "Yes."

Caroline: "From what I remember, the box contained the evils of the world, released when Pandora opened it. However, it also contained hope, which remained inside. My English teacher said the box represents our curiosity and desire for knowledge, and that obtaining knowledge can and will have both negative and positive consequences. He also pointed out similarities with the story of Adam and Eve: the pursuit of knowledge with dangerous consequences. Although both stories explore the tension between the desire for knowledge and the potential for harm, they differ in one very important respect: in the Greek myth, the presence of hope inside the box, regardless of whether it is seen as a blessing or another form of torment, foresees with extraordinary prescience the state of the world, with the amoral tendencies of geopolitics dominating human behavior.

Don then showed Caroline a plot he produced as an undergraduate, which illustrated the rate at which nuclear weapons had spread since the advent of the atomic bomb.

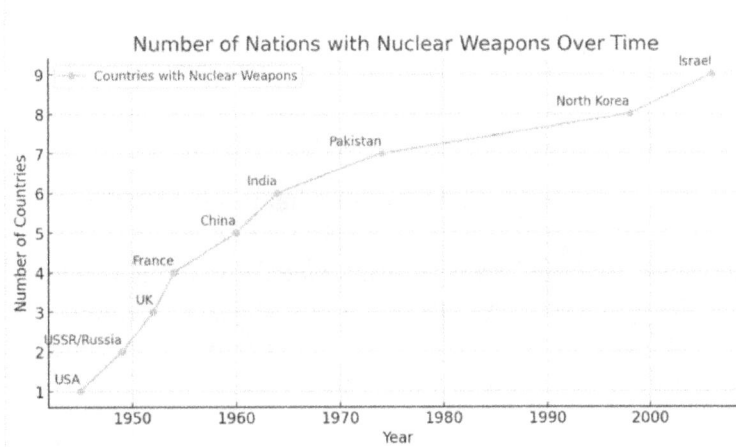

Figure 3. The number of nations with nuclear weapons continues to increase

Don: "In addition to the nine countries that possess nuclear weapons, five of the countries that are members of NATO participate in nuclear-sharing agreements with the United States, meaning they host U.S. nuclear weapons but do not possess their own."

Caroline: "That's really frightening. The only thing I heard from Dr. Siraganian that seemed reassuring was that in 1972 the United States and the Soviet Union signed a treaty limiting both of them to 22 anti-ballistic missile sites each. But then in 2001, the United States withdrew from the treaty, arguing that new threats, including terrorism and rogue states, required modern missile defenses. Russia opposed the decision, warning it could lead to a new arms race."

Don: "Western nations argue that Russian fears have not materialized, and perhaps that's true. On the other hand, withdrawal removed a significant constraint on the development and deployment of missile defense systems by the U.S., Russia, China, and other nations, leading to a potential arms race in defensive technologies. Then, of course, there's the intangible but perhaps even more important erosion of trust, which makes it more difficult to negotiate further arms control agreements—even though efforts are still being made."

Caroline: "So the genie is out of the box—or Pandora as the Greeks would have it—and it looks like there's no way to get it back inside."

Don: "Yes, that's the way it seems. Technological events tend to be irreversible, and their consequences are very difficult to control. It's like entropy in the physical Universe—it moves only in one direction, and to keep it from increasing requires a great amount of energy.

"I can't help recalling words from the Rubáiyát of the great Persian mathematician and poet Omar Khayyám."

The Moving Finger writes: and, having writ,
Moves on: nor all thy Piety nor Wit
Shall lure it back to cancel half a Line,
Nor all thy Tears wash out a Word of it."

Caroline: "The words are both beautiful and astute. But what is a Rubaiyat? And what's entropy?"

Mary: "A rubaiyat is a collection of four-line poetic stanzas called quatrains. Omar Khayyám wrote hundreds of them..." Then Mary added in a jocular manner which indicated how awed she was by his breadth: "probably in his off moments between classifying cubic equations and finding geometric solutions, while exploring Euclid's parallel postulate."

Mary, looking at Don: "Do you want to take a shot at explaining entropy?"

Don: "I'll try. I think it's easiest to explain by example. Consider a room filled with air and completely closed off. The air molecules move about randomly, every which way. A scientist would say they're in a state of maximum disorder. Now suppose we had a big piston which moved all the molecules to one half of the room, the other half being a vacuum. That would obviously require the input of energy, but as soon as the piston was removed, what would happen?"

Caroline: "The molecules would redistribute and fill the entire room again."

Don: "Yes, the molecules would rearrange themselves spontaneously, moving from a partially ordered state in which they filled only half the room to a totally disordered state. *Entropy is a measure of disorder*, and in that isolated room, entropy changes only in one direction. Think of Lewis Carroll's catchy nursery rhyme, Humpty Dumpty—there was no way to unbreak the egg; just as there is no way to put the genie back in the bottle, or for the molecules to move by themselves to just one side of the room."

Caroline: "That's interesting."

Don: "What I've described is a version of the Second Law of Thermodynamics. There's just one small caveat in this description, which you'll learn about in college when you take a subject called statistical mechanics: it's statistically possible for the molecules to move spontaneously to one half of the room, but it's extraordinarily unlikely—for all practical purposes, the probability of it happening can be considered to be zero."

Caroline, now moving the conversation back to the world of Planck and Einstein: "Dad, did you know that Max Planck had something important in common with you?"

Don, with uncertainty in his voice: "We both love physics?"

Caroline: "Well, yes, but that's not what I had in mind. He was a talented musician and was torn between a career in physics and a career in music. But of course, unlike you, he chose physics."

Don: "That's interesting, and something I never knew. I wonder what would have happened in an alternate Universe where I chose physics, and he chose music."

Caroline: "Alternate Universe?"

Don: "Remind me to tell you about the many-worlds interpretation of quantum mechanics sometime—or better yet, ask your uncle. It might sound like science fiction, but it's deeply rooted in quantum theory, though I'm not sure how many physicists take it seriously."

Caroline: "That sounds weird, but I guess quantum physics for me is still a few years away. For now, I have a question, though it's more about history than science."

Don takes a deep breath, wondering if he'll be able to answer: "Sure, what's the question?"

Caroline: "Max Planck was one of the most brilliant physicists of his generation. He knew that visible light was a form of electromagnetic radiation, and he must have known that the photoelectric phenomenon couldn't be explained by the wave theory of light—and he was the one who introduced the idea of quanta. Yet he refused to accept a quantum hypothesis as a possible explanation for the photoelectric effect."

Don: "Yes, that is remarkable. There's no evidence that Planck accepted a connection between his quantum hypothesis—which, by the way, he considered a mathematical trick more than a hypothesis—and the photoelectric effect. This highlights the extraordinarily conservative nature of science and the reluctance of even the most brilliant minds to let go of, in Kuhn's words, an existing paradigm. It wasn't until five years after Planck introduced the idea of quantized radiation that the next generation's brightest mind—Einstein, who was more than two decades younger than Planck and was eventually hired by him as a professor of physics—had the courage to take a bold step and revolutionize the way physicists understood light."

X-rays Can Display Anatomy Because They Are Absorbed More Effectively by Bone Calcium Than by Soft Tissue

Caroline, with a smirk: "Yes, I get it—Einstein was able to see the light. Anyway, even though I was initially surprised by Uncle Zack's digression and thought he was going off track, his explanation of the photoelectric effect helped me understand the X-ray patterns Roentgen observed. In the case of bone, an electron is ejected from calcium, whereas in the traditional photoelectric effect, electrons are ejected from atoms on a metal surface."

Don: "I believe the key to understanding bone's relatively high efficiency in blocking X-rays is that calcium, with its 20 protons, is heavier than hydrogen, oxygen, and nitrogen, the most common atoms in soft tissue."

Caroline: "Yes, Uncle Zack went even deeper—he said that the probability of an X-ray ejecting an electron from an atom is proportional to the cube of the atom's atomic number. This means that bone is much more effective at blocking X-rays than soft tissue.

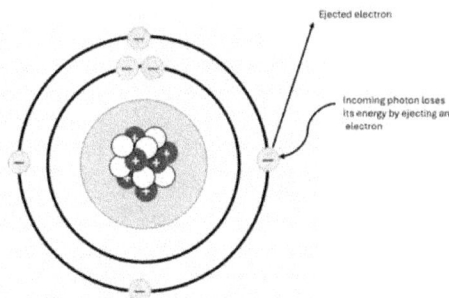

Figure 4. When a photon's energy is greater than the energy with which an electron binds to its atomic nucleus it can knock the electron from its orbit. The free electron will eventually dissipate its energy as heat, possibly after colliding with other atoms.

"The explanation was helpful, but it didn't leave me completely satisfied—it raised as many questions as it answered, such as why the probability of an electron being ejected from an atom depends on the cube of its atomic number; what happens to the ejected electrons; and finally, since soft tissue doesn't contain calcium, what properties enable it to absorb X-rays?

"Regarding the first question, Uncle Zack explained that heavier nuclei have more protons and are, therefore, more positively charged than lighter nuclei—and that allows them to bind electrons more tightly. Since ejecting a tightly bound electron requires more energy than ejecting one that is loosely bound, the larger the atom, the more effective it is at blocking radiation. He also mentioned that the reason the probability of ejection depends on the cube of the atomic number, rather than some other power, could only be explained using quantum mechanics, so I'd have to wait for that explanation until I took atomic physics.

"I also asked Uncle Zack what happens to the ejected electrons. Instead of explaining, he said that I now knew enough to research the process, which is called photoelectric absorption, on my own. On the other hand, he gave me the bottom line: the energy of an incoming photon that never reaches the photographic plate is eventually dissipated as heat. However, there can be multiple steps in the process, and I would need to find out what they were."

Mary: "It sounds like you had quite a conversation."

Caroline: "Yes, a lot to take in—but there was also something else, non-scientific, which seemed a little puzzling."

Don: "Oh?"

Caroline: "When I asked how soft tissues block X-rays and why they do so less effectively than bone, Uncle Zack held a Zoom meeting with Aunt Zoe, and when they first connected, and when they looked at each other, I had the strangest feeling—they seemed genuinely happy to see each other, yet there was also a sense of sadness, almost longing. It's hard to describe. In any case, together they gave me what, in retrospect, was an obvious explanation."

Mary: "We're listening."

Caroline: "Density, of course, matters, but the attenuation process is essentially the same as in bone—X-rays induce electron ejection from heavier atoms, mainly oxygen and nitrogen, which have atomic numbers of 7 and 8, respectively. DNA also contains phosphorus, which has an atomic number of 15. None of these atoms is as heavy as calcium, which has an atomic number of 20, so the abundance of calcium in bone is the main reason that it blocks X-rays more effectively than soft tissue. Uncle Zack also pointed out that resonance could be ruled out as a reason for X-ray attenuation because typical bond frequencies are several orders of magnitude lower than the frequency of X-rays."

Don responded with a seemingly unrelated question: "Did your mother and I ever tell you how we met?"

Caroline: "No, how did you meet? And why are you asking me that now?"

Don: "We can talk about it later; it's an interesting story."

Though Caroline was intrigued, she had little difficulty letting go of her father's puzzling question and continued the conversation. "I also learned a little about DNA. When I told Aunt Zoe and Uncle Zack that I had no idea DNA contained phosphorus, Aunt Zoe explained that a DNA strand consists of a series of phosphates, sugars, and bases called nucleotides. She then showed me a cartoon representing a sequence of nucleotides. I had, of course, heard of the DNA double helix, even though I haven't yet taken biology, but I had never seen an atomic-level representation."

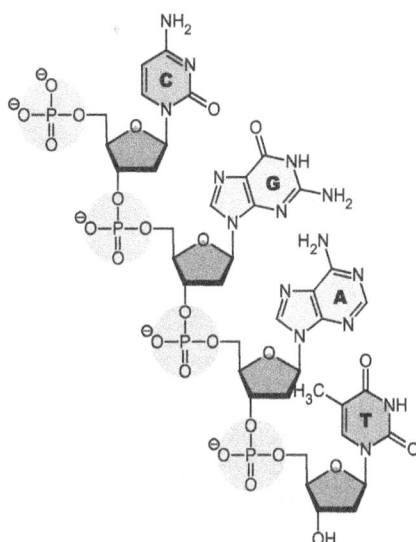

Figure 5. A four nucleotide segment of a single strand of DNA, consisting of phosphate groups (a phosphorous atom bonded to 4 oxygen atoms, green), sugars (gray) and bases (labeled C,G,A and T).

The conversation had both broadened and deepened Caroline's curiosity. The explanation of why bone is less transparent to X-rays than soft tissue led her to the more general question of why certain substances are transparent while others aren't. She wondered, for example, why visible light cannot pass through wood or metal or many other materials which are, after all, mostly empty space, while it can pass through glass and diamond, among other hard substances. At the same time, what she had learned about DNA made her wonder how scientists know the atomic structure of molecules in such detail.

Then she had a strange thought: She had just learned that Roentgen was able to photograph his wife's hand using X-rays, and she wondered—though she was almost afraid to ask—whether X-rays could also be used to photograph DNA and other large molecules. But uncharacteristically, she let the question slip because she couldn't even come close to imagining how that might be done.

It was Friday evening, and after a grueling week filled with exams and deadlines, Caroline was more than ready to unwind for the weekend. She had a few simple plans: practice her violin, catch a movie with friends, and—perhaps most importantly—dive deeper into *Time and Again*, an absolutely captivating novel by Jack Finney. She couldn't quite pinpoint what it was that had her so hooked, but one thing was clear: she wasn't alone in her fascination. The legendary astronomer Carl Sagan had praised the book for being "so tautly constructed, so rich in the accommodating details of an unfamiliar society," that he was "swept along before [he] had a chance to be critical."

Even Stephen King, a master of fantasy storytelling, called it one of the greatest time-travel novels ever written, a glowing endorsement from one of the genre's most esteemed voices.

Figure 6. Dall-E's rendition of New York's Central Park in winter. The view is west toward the Dakota apartment building with the New York Society for Ethical Culture behind it, and a fictitious planetarium across the street. This could well have been a portrait from the late 1800s, except for the anachronistic skyscrapers.

To Caroline, *Time and Again* was more than just a story set in an unfamiliar society; it was about her beloved city, presented as a romanticized version of its adolescence now lost to time—a city revealed through the poignant sketches of a time traveler. These sketches depicted the newly constructed Dakota apartment building with its magnificent French-German architecture; picturesque, unhurried horse-drawn trolleys; haunting glimpses of New Yorkers strolling serenely through snow-covered Central Park in winter; a peaceful, long-gone Madison Avenue with Vanderbilt's brand-new mansion, made of dazzling white limestone; and the Third Avenue El train in a bright, airy environment, vastly different from its dark and dreary remains shortly before its demolition. Alongside this romantic vision was a love story—of a protagonist traveling across time to fall in love with a woman who seemed to exist only in his imagination.

But then, in the middle of her musings, the phone rang. It was Noreen, calling because she couldn't stop thinking about a problem their biology teacher had assigned for homework: How far would the DNA from all human genes stretch if laid end to end? It wasn't so much that Noreen couldn't answer the question—it was that, after calculating the distance between adjacent base pairs (3.5×10^{-10} m) and multiplying it by the approximately 3 billion base pairs, she had come up with a distance of 1 meter. The problem was that the diameter of a human cell is roughly 170,000 times smaller than this calculated distance. And it wasn't just the incredible ratio that struck her—it was

134

the real difficulty of understanding how so much DNA could fit into such a small space without becoming horribly entangled.

Noreen: "I can't even begin to visualize how such a long strand could be packaged in an orderly way in such a small space."

Caroline: "Neither can I, and it's especially puzzling when you think about how buried genes can interact with the large proteins needed for transcribing DNA into RNA."

Noreen: "I guess there must be a way to open the package."

Caroline: "Yes, but I have no idea what the package looks like. There must be methods for determining its structure, but now that I think about it, I don't even know how the structure of DNA was determined. Everyone says it's a double helix, but what's the evidence?"

Noreen: "Dr. Kahn said that Watson and Crick, a biologist and a physicist who collaborated closely, used X-ray diffraction to help unravel the structure, but I don't understand what that means. Diffraction occurs when light encounters an obstacle and spreads around it, as though it bends But I can't imagine how that happens. I never thought of light as something that can bend. I also don't understand why they had to use X-rays to determine the structure. Why couldn't they just see DNA through a high-magnification microscope that amplifies light in the visible part of the electromagnetic spectrum, or use some other type of radiation?"

Caroline: "I'll ask my father; maybe he can explain diffraction. If he can't, I'll ask Uncle Zack— but not right away; I really need this weekend to relax. Plus, *I just learned Einstein's theory of photoelectricity, which says that light behaves as a particle, and now we're talking about diffraction, which is a property of waves.* I'm completely puzzled, and hope Uncle Zack can explain how it can be both."

Noreen was disappointed, but she and Caroline agreed to continue discussing DNA and X-ray diffraction over lunch on Monday.

The next morning, Caroline and her parents had barely sat down to breakfast when she asked how scientists know that DNA has the structure of two intertwined helices—and, as expected, they referred her to her aunt and uncle. Though Caroline had promised herself a weekend free from science, she called her aunt immediately after breakfast and could barely wait to speak with her uncle—she only held off until noon because of the two-hour time difference between New York and New Mexico.

By late afternoon, Caroline had spoken with both of them and organized her notes, though she said little to her parents about the conversation. She only mentioned that she needed time to assimilate what she had learned and that she would need a follow-up discussion with Uncle Zack because he had to cut the conversation short to attend an important meeting on fusion energy. Then it was off to a movie with Noreen and other friends. Finally, on Sunday, Caroline enjoyed the much-needed relaxation of playing her violin and reading the newspaper that's famous for crossword puzzles.

Caroline and Noreen spent most of the next week reviewing her notes and going over what they knew about light waves, which are often drawn as sinusoidal curves to represent their intensity, or amplitude, variation. They knew that when two waves with the same maximum amplitude meet and their peaks align perfectly, the resulting amplitude is twice as large as that of the individual waves.

In algebraic form, if $A_1 = A \cos \theta$ and $A_2 = A \cos \theta$, then their sum is:

$A_T = A_1 + A_2 = 2A \cos \theta$.

For a physical wave such as light or sound, $\theta = \omega t$, where ω is 2π times the vibrational frequency. In other words, if f is the vibrational frequency, then $\omega = 2\pi f$.

In this simple situation with coinciding peaks and equal wavelengths, the waves are said to be *in phase*.

Caroline and Noreen were curious about the result of adding waves that are *not* in phase. In other words:

$A_T = A [\cos(\theta + \varphi) + \cos \theta]$,

where φ is the phase difference—the number of radians (or degrees) by which one wave leads or lags the other. They remembered an identity they had learned just recently in their intermediate algebra class which related the sum of two trigonometric functions to a single function.

$\cos a + \cos b = 2 \cos(\tfrac{1}{2}(a + b)) \cos(\tfrac{1}{2}(a - b))$

It was just a matter of substituting $a = \theta + \varphi$ and $b = \theta$ to obtain:

$S \equiv A_T / A = 2 \cos(\varphi/2) \cos(\theta + \varphi/2)$

They were stunned by the result: the ratio of the amplitude of the combined waves to the original wave is twice the cosine of the average phase shift, $(\varphi + 0)/2$. They realized immediately how much information this relation summarized—if the phase shift is π, then $S = 0$, and interference is completely destructive; if the phase shift is 4π, the amplitude doubles and interference is completely constructive. More generally, for any whole number *n* (i.e., n = 0, 1, 2, 3, etc.), interference is completely destructive if $\varphi = (2n + 1)\pi$, and completely constructive if $\varphi = 4n\pi$.

For other phase delays, the equation enables an immediate determination of the resulting amplitude. For example, if the phase shift $\varphi = \pi/2$, then the expression indicates that the resulting wave has an amplitude of $\cos(\pi/4) = \sqrt{2}/2$, a phase that's half that of the original phase difference between the two waves, and a frequency that's unchanged. Such is the power of mathematics—that a simple expression can convey so much information (Figure 7).

Caroline and Noreen learned that light waves are composed of oscillating electric and magnetic fields, which are generated by moving electric charges. And the intensity of light—its energy per unit time per unit area—is proportional to the square of its amplitude. While they both were fascinated by the fact that intensity varied quadratically with the field strength, there were still too many gaps in their understanding of electromagnetism for them to fully grasp why this relationship exists. For now, though, they were content to understand that what we perceive as brightness corresponds to the wave's intensity: the rate at which visible light energy reaches the human retina, divided by its area.

Perhaps most astonishingly, Zoe informed her niece that the threshold of the human visual system—the minimum intensity of light required for detection—can be as low as one trillionth of a watt per square meter. Given that the human retina covers an area of about one millionth of a square meter, this means that the human eye is extraordinarily sensitive, capable of detecting light at power levels as low as one attowatt, i.e., one billionth of a billionth of a watt.

By the next weekend, Caroline felt ready to share with her parents what she had learned, and she invited Noreen to join the conversation.

"Uncle Zack and Aunt Zoe both have a lot going on, but we did discuss some of the principles behind the technologies used to determine molecular structures. I took some notes, which Noreen and I reviewed together last week. They explained that X-rays can be used to help infer the structures of proteins and DNA from diffraction patterns, but neither Noreen nor I know much about diffraction, except that it somehow involves the spreading of a light beam after it passes through a small opening.

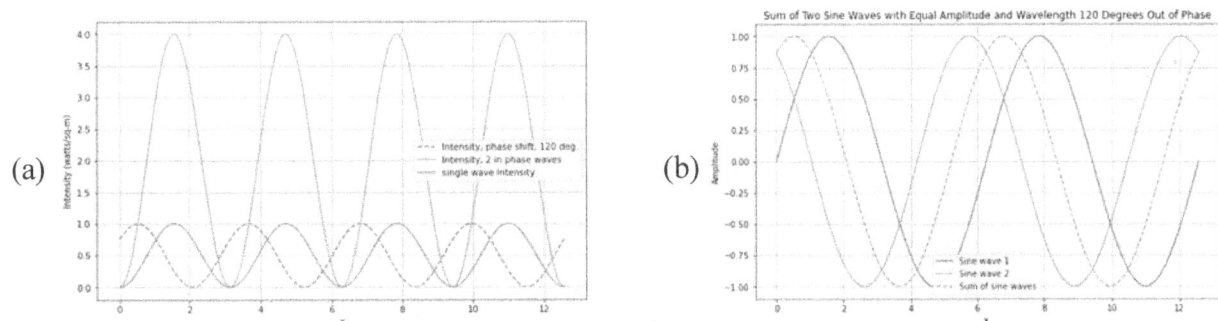

Figure 7. (a) Two waves of unit amplitude and 120° out of phase interfere so that the combined wave also has an amplitude of 1 and is phase shifted 60°. (b) Since the intensity is proportional to the square of the wave amplitude, the intensity that results from adding 2 waves that are in phase is 4-fold higher than that of the single wave.

"Uncle Zack began by describing an experiment in which a laser that emits light of wavelength λ is directed at a barrier with a tiny slit, the width of which is comparable to the wavelength— meaning $d \sim \lambda$. When the laser illuminates the barrier, a screen which is behind it, displays a remarkable and entirely unexpected symmetric pattern: a bright central peak, flanked by peaks whose brightness diminishes as the distance from the center increases."

Don: "Yes, that is remarkable — who would possibly have thought ... ? Where did that pattern come from? Did he explain how it can be interpreted in terms of the wave theory of light."

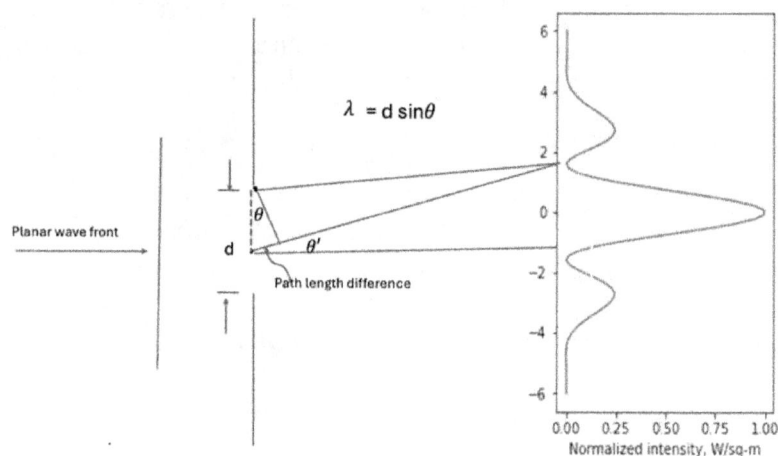

$$\lambda = d \sin\theta$$

Figure 8. If light were corpuscular only a single spot would appear on the screen, rather than the pattern shown which is consistent with interfering waves. Huygens's principle (Fig 9) says that each point on a wave segment that enters the slit, two of which are shown, one at the edge and one at the center, will generate a wavelet. In the figure the emerging wavelets are not shown, but the directions of two of them are represented as straight lines terminating in the first minimum of the diffraction pattern. The brightness of any point on the screen depends on the phase relation between the wavelets that arrive there and the distance that they have traveled. The first minimum is produced by wavelets that have a path difference equal to half a wavelength. They are therefore 180° out of phase and interfere destructively. The angles θ' (which is easily measurable) and θ (which is not) are essentially equal.

Caroline: "Yes, he tried, but I didn't fully grasp it. Apparently it can be understood in terms of a principle first proposed by the Dutch physicist Christiaan Huygens.

"As an aside—and I enjoy your brother's occasional asides—he said that Huygens was a polymath. And I said, "a what?" He didn't define the word, he just said that Huygens, in addition to making important contributions to physics, made major contributions to mathematics, astronomy, and engineering. And to top it off he also invented, among other things, the pendulum clock, which served for centuries as the standard for keeping time—so even without a definition I got the picture.

"*Huygens hypothesized that each point on the surface of a light wave is itself the source of a tiny wave, a so-called wavelet, which is in phase with its source.* Because all wavelets are in phase, they interfere constructively so that their envelope, i.e., the curve tangent to all the wavelet peaks, is the source wave an instant later.

138

Wavelets are emitted by every point on S_1. Wave front formed by wavelets

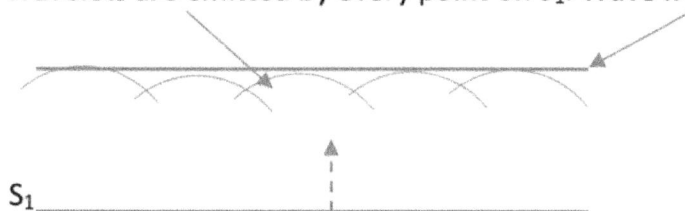

S_1

S_1 represents a segment of a wave emitted by a distant source

Figure 9. As a spherical wave leaves its source it expands, and every segment of its surface becomes a smaller and smaller fraction of the total area so that it eventually looks planar. According to Huygens a wave progresses by forming in phase wavelets.

"He then explained how Huygens's principle accounted for the surprising results of a simple experiment in which light is directed at a barrier with a small slit. When a wave reaches the barrier, it is blocked except for the small segment that passes through the slit. If light were purely particle-like, one would expect to see a single spot on the screen behind the barrier. Instead, a pattern consistent with wave interference appears.

"To explain this pattern, Uncle Zack sketched two light rays, separated by d/2 within a single slit and showed that they can interfere destructively. For complete cancellation, their path difference must equal half a wavelength. For the first minimum, two rays separated by d/2 cancel if their path difference is ½λ. Generalizing this argument, the condition for destructive interference minima in a single-slit diffraction pattern is

$$d \sin \theta = m\lambda, \qquad m = 1, 2, 3, \ldots$$

Uncle Zack also mentioned that these relationships rely on θ being equal to θ'—but instead of proving it, he asked me to do it, and I have no idea where to begin.

"At that point, we had to end the conversation because he had a meeting with colleagues to discuss recent progress in fusion energy."

Don: "Energy from fusion? That will have a monumental impact on civilization if it works."

Caroline: "Yes, he's very excited about the possibility and working very hard to help make it happen. He said that harnessing nuclear fusion could, in principle, provide an endless and inexpensive energy source for the entire planet, revolutionizing both individual lives and international relations. However, progress had been slow. He mentioned that recent work at Lawrence Livermore Lab, Los Alamos's sister lab, had achieved 'ignition,' meaning that the energy released from fusing tritium and deuterium was, for a brief moment, greater than the energy input from the 192 lasers used to trigger the reaction. While that was a breakthrough, when considering the total input energy required to power the lasers, the output still remained below the energy input."

Don: "There's a standing joke: fusion is always only a few decades away—in other words, nothing substantial is happening. But now apparently that's beginning to change—the recent developments at Lawrence Livermore and Los Alamos give real hope that this time a 30-year projection might actually be accurate."

Caroline: "Let's hope so. But to return to something more straightforward, Uncle Zack briefly described a famous experiment performed by Thomas Young, similar to the single-slit experiment but with two slits in the barrier. The concepts are similar to those in the single-slit experiment. When the paths of two light rays—one from the upper slit and one from the lower—converge at a bright spot on the screen, the interference is fully constructive, and the path difference Δ is a multiple of the wavelength. This means a bright spot occurs whenever $d\ sin\theta = n\lambda$. Although θ isn't directly measurable, it essentially corresponds to the measurable angle θ', i.e. ∡DOP.

Thomas Young's Double Slit Experiment Established the Wave Nature of Light

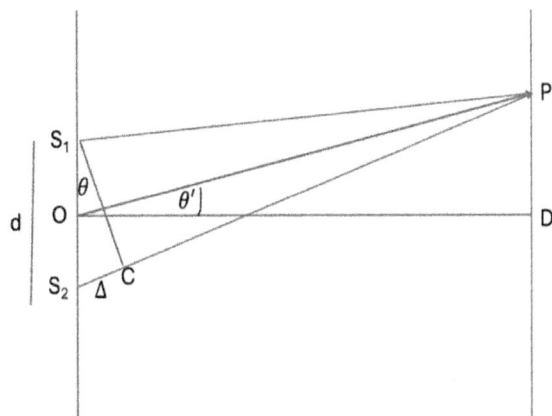

Figure 10. Young's Double Slit Experiment. S_1C is perpendicular to S_2P and cuts it into two segments, one of which is equal in length to S_1P, and the other, by definition, is the difference Δ in their path lengths. Wavelets from the two slits will meet at a bright spot P on the screen if Δ is an integer multiple of their wavelength.

"In between meetings, I did try to prove that $\theta' = \theta$, but I didn't get very far. Uncle Zack then said, somewhat apologetically, that "perhaps I had misled you—the angles aren't exactly the same, but the error in using them interchangeably is negligible if the distance from the barrier is very large compared to the distance between the slits." He also pointed out something I had never realized: when tackling a physical problem, it's useful to closely examine the sizes of any objects or variables involved. In this case, these would be the distance between the slits, the distance from the slits to the screen, and the wavelength of light. Their relative magnitudes can suggest useful approximations that turn an otherwise difficult problem into one that's easily solvable.

"When I thought about it and started drawing diagrams of the situation, it didn't take long to realize that if the ratio of the slit distance to the screen distance is very large, the light rays emerging from the slits will be virtually parallel—at least for a while."

Caroline then opened her notebook and showed how she had solved the problem.

Don: "How good is that approximation?"

Caroline: "I did some reading and found that reasonable values for the distance between slits and the distance to the screen are $d = 1300$ nanometers (nm) and $L = 3$ m, so $d/L = 4.3 \times 10^{-7}$. That means at distances relevant to the approximation, a few tens of wavelengths from the slits, or even hundreds of wavelengths from the slits, the waves are essentially parallel—they converge extremely slowly. With blue light, for example ($\lambda = 450$ nm), waves that are 1300 nm apart at the slits will be (1300 - 0.02) nm apart after they've moved a distance of 100 wavelengths, and even after they've moved 10,000 wavelengths, they will still be (1300 - 2) nm apart. In the first case, they will have deviated 0.0015% from their initial separation, and in the second case, 0.15%—so the error at distances relevant to the approximation is negligible."

Her uncle also encouraged her to think about a related topic: the distance between two objects at which they just become unresolvable—or, stated as a question: what is the criterion for concluding that you are looking at two objects rather than one? Such a condition was proposed by Lord Rayleigh, who stated that "two objects are no longer distinguishable when the first interference minimum of one object coincides with the central maximum of the other." In other words, the separation of the peaks subtends an angle θ_1. Consequently, the rule of thumb is that to resolve two objects separated by a distance d, the wavelength must satisfy $\lambda = d \sin \theta_1$; smaller distances require smaller wavelengths. This relationship can also be expressed in terms of the spatial distance on the screen between the two central maxima for small θ_1. Specifically, if y_1 is the distance on the screen between the central maximum and the first minimum, then, assuming the distance to the screen L is large and θ_1 is small, $y_1 \approx L\lambda/d$.

In a typical double-slit experiment, $d = 0.0001$ m, $1\ m \leq L \leq 2\ m$. For blue light with a wavelength $\lambda \approx 450$ nm, a typical distance on the screen between the central maximum and the first minimum is $y_1 \approx (1.5 \times 450 \times 10^{-9})/10^{-4} = 0.06$ m, so it's roughly 2 to 3 inches, and the spot is in any case fuzzy, so the distance between spots can only be approximated.

Light Behaves as Both Wave and Particle

Caroline then returned to Huygens's theory, which explained the observations in terms of waves. "The idea that a wavefront emits wavelets that interfere with one another to form the progressing wave seems little more than a device—it seems unnatural, made up to explain diffraction."

Don: "You're right. My understanding is that the modern view explains diffraction and interference quantum mechanically. I don't understand the quantum mechanical view in detail, so you should ask your uncle about it—but even that has aspects that seem unnatural."

Caroline manages to wait until the weekend to call her uncle.

Zack: "Hi, Caroline, I'm ready for this morning's grilling."

Caroline: "I'm sorry to keep bothering you, Uncle Zack. I really appreciate the time you've taken to explain physics, and I know you're involved in a bunch of important projects, but I'm thoroughly confused."

Zack: "It's my pleasure, niece—I enjoy our conversations, and I certainly find them challenging."

Caroline: "Here's the thing: Einstein's theory of photoelectricity proves that light is quantized and consists of photons, which are discrete particles. On the other hand, the double-slit experiment shows definitively that light is a wave. So, which is it?"

Zack: "It gets even more confusing. If you use a laser that emits photons very slowly so that they're well separated and can't interfere with one another. As each one arrives at the screen, you'll see its signature, which means that at first, you'll see scattered dots. But after a while, a pattern will begin to form, and it's precisely the interference pattern shown in Young's experiment. The implication is that each photon interferes with itself, or as physicists say, the wave function interferes with itself—i.e., the photon passes through both slits simultaneously. That, of course, is entirely in conflict with common sense. So you say, '*Let's do an experiment that will reveal which slit each photon goes through.' The result is that the interference pattern disappears and is replaced by two distinct clusters of spots*. In other words, you're free to hypothesize something apparently ridiculous—in this case, that *a photon goes through both slits simultaneously—as long as there's no way to falsify it. Physicists refer to this strange behavior as wave-particle duality.*"

Caroline: "That's mind boggling; it's unreal."

Zack: "I know. It might help to think of light as both a ripple on water and a tiny particle, depending on how you observe it. *In the double-slit experiment, light spreads out like a wave and creates an interference pattern, but when we try to track its path, it behaves like particles, and the pattern disappears.*

"No one truly understands it. A speculative proposal called the 'many worlds' interpretation tries to resolve the difficulty by hypothesizing that whenever a branch point occurs in the world, such as a photon having to go through one or another slit, the Universe splits. The photon goes through different slits in different universes—so it doesn't go through both slits simultaneously. But the cost is the creation of an infinite number of universes. Every outcome of any event is realized in some universe."

Caroline: "It sounds like science fiction."

Don: "So, if there are an infinite number of universes, maybe in one of them I'm a quantum physicist answering your questions!"

Mary: "Or maybe in another Universe, Uncle Zack is the one asking you the questions!"

Zack: "Since your father was on his way to becoming a physicist, perhaps that's not so strange. Who knows? In any case, good science fiction often tells stories that can happen—it doesn't violate physical laws, but the events are not currently technologically possible and may not be for the foreseeable future. I don't think any science fiction story has ever anticipated a law or principle that was later discovered; science fiction isn't imaginative enough to reveal the incredible mysteries of the Universe. That idea about the limits of human imagination is perhaps best embodied by the words of one of the most perspicacious people ever to walk the face of our planet:

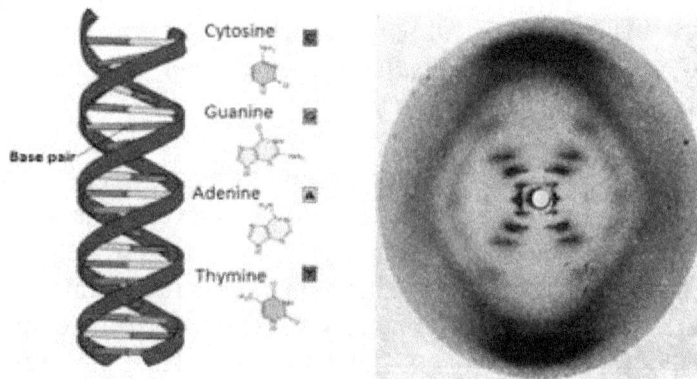

Figure 11. A cartoon of the DNA double helix (https://brainly.com/question/8998157), left and the diffraction pattern from which information about its structure can be obtained. Strikingly, unlike an X-ray of a macroscopic object such as a human organ, the cartoon on the left, which captures the classic DNA double helix, looks nothing like the diffraction pattern on the right.

There are more things in heaven and earth, Horatio,
Than are dreamt of in your philosophy.
—*Hamlet*, Act 1, Scene 5, William Shakespeare."

X-rays Can Help Visualize Proteins and DNA

Caroline: "Thank you, Uncle Zack. I guess it doesn't matter much whether light sometimes behaves as a particle or a wave—what matters is that *the diffraction of light can be used to infer the structure of molecules*. Now I might be able to understand something about how that's done."

Zack: "You should speak with your Aunt Zoe about that."

The following Saturday Caroline does as her uncle suggested and has a phone conversation with her aunt.

Zoe starts by explaining that the structure of DNA was not determined by X-ray diffraction alone. "There is a great deal of evidence from biochemical studies and physical models suggesting that DNA is helical, but that evidence isn't sufficient to determine the precise locations of its components, such as the bases. What Watson and Crick showed was that the diffraction patterns produced by DNA were entirely consistent with it having a double-helix geometry."

Caroline: "What I found striking is that the DNA diffraction pattern Zoe showed me looks nothing like the familiar double helix. Using x-rays to determine molecular structure is very different from the way they are used in medicine to detect abnormalities. The procedure used in medicine seems to be more or less what Roentgen did: x-rays are passed through an organ that is suspected to be abnormal, and the image is compared with that of a healthy organ. In science, advanced mathematics is required to determine the structure. Uncle Zack said the difference is that in medicine, the image is produced by differential absorption of the radiation, whereas in biology, it is determined by an analysis of the diffraction of radiation.

"Uncle Zack, I think just to be cute, showed me via Zoom, one of the equations that

Crick and his colleagues needed to derive

$$S = \sum_{s=0}^{N-1} \exp\left[i2\pi Rr \left(\cos\psi \, \cos\frac{2\pi s}{N} + \sin\psi \, \sin\frac{2\pi s}{N} \right) + \frac{i2\pi \zeta sc}{N} \right].$$

"The equation is way beyond me, and he didn't try to explain it—he just wanted to make the point that determining the structure wasn't just a matter of looking with your eyes."

The Electronic Structures of Glass and Wood Explains Why Light Can Penetrate One But Not the Other

Caroline felt that she had reached the limit of her understanding about light, a strange phenomenon that was so common and yet so mysterious. But she was still puzzled by something: why can't people see through so many common substances, like wood and metal, while being able to see through others, like glass and diamond? At first, she thought some kind of resonance might explain why visible light didn't penetrate wood, and she dismissed the idea of photoelectron ejection because visible light has much lower energy—about four orders of magnitude lower than X-rays.

However, Caroline felt a flicker of understanding emerge when she came across information about two of the macromolecules that are major components of wood: lignin and cellulose. Lignin can absorb visible light because some of its bonds vibrate in the optical region, and fibers of cellulose are similar in size to the wavelength of visible light, making them scattering centers. In addition, some photons don't penetrate wood at all; instead, they're reflected and scattered by the bark.

Caroline was beginning to develop a basic understanding of how light interacts with matter, even though she still had no grasp of how light could be both a wave and a particle. But one big question still loomed: why is glass transparent to visible light but not to other parts of the electromagnetic spectrum?

Judging by what she had learned so far, she thought the answer might involve the molecular structure of glass, the wavelengths of incoming photons, and either the vibrational frequencies of the molecular bonds or the structure of the electron orbitals.

Her father had told her that glass is made of silicon dioxide (SiO_2), but how that made it transparent wasn't clear to her. "Oh dear" she thought, realising that clear had a double meaning, "I'm starting to think in puns the way my father does."

She decided to call her uncle for help. He explained that one of the main reasons glass is transparent is that it doesn't have free electrons that can collide with incoming photons. Additionally, the electrons bound to the atoms in glass are very stable and knocking them loose to transfer energy requires higher-energy photons than those in the optical range. To put it in chemist's terms, there is a large energy band gap between the free and bound electrons, which makes glass an insulator.

Curious, Caroline asked why other insulators aren't also transparent. Her uncle thought it was a great question and explained that glass's transparency has to do with the fact that it's amorphous— it has no regular arrangement of molecules. Caroline, always keen for more answers, asked why being amorphous helps facilitate transparency. Zack suggested that she think back to what she'd learned about wave interference: waves can only interfere when they scatter from regular atomic arrangements. If there is no regular arrangement, the waves scatter in all directions—including the forward direction.

Caroline now understood why glass is transparent, and she had that wonderful feeling of comprehension that comes with figuring something out for the first time. It was all the more gratifying because she had arrived at the answer by asking the right questions. But one final question remained.

Just after dinner one evening, as her parents moved to the living room to relax, Caroline lingered instead of heading to her room to do homework. She said, somewhat haltingly, "Now that I understand why some materials are transparent, I'm wondering about another common property of light: reflection."

She was pleasantly surprised when her father, instead of telling her to wait for the weekend, seemed genuinely receptive.

Don: "Great question! Reflection depends on the way a material's surface interacts with light. A glass mirror has one side coated with a thin layer of metal, like aluminum or silver. These metals have lots of electrons that aren't tightly bound to atoms, which means they can move around easily. When light hits the surface, the electric part of the wave pushes on the electrons, causing them to oscillate. The oscillations produce their own electromagnetic waves, and some of the waves travel back in the same direction as the incoming light. That's what we see as a reflection. The entire process is based on how the light's electric field interacts with the electrons and the smooth surface of the metal."

Conversation 7

Our Planet's Changing Climate

Caroline is now a high school junior and has been hearing a lot about climate change—and naturally she has countless questions. As it happens her Aunt Zoe has recently become engaged to Barry Bruce, a geophysicist who knows a great deal about Earth's climatic history and its current condition. Caroline meets him for the first time when her parents invite Zoe and Barry to dinner, and as expected he welcomes Caroline's questions. The five of them engage in a stimulating conversation covering various topics: the evidence for a changing climate, the proof that these changes are anthropogenically driven, how the current changes relate to the more dramatic fluctuations in Earth's history, the significance of these changes, what can be done to address them, and why so little is being done. Although Barry is initially taken aback by Caroline's brash inquisitiveness, he is also impressed by her curiosity and the evening ends, to the delight of her parents and aunt, with Caroline in absolute awe of her 'uncle to be'.

The Earth is an Organism—and it Has a Fever

It's Sunday morning, just after breakfast. The Angstroms are gathered in their family room. Caroline, who had a fun-filled Saturday, is now feeling the pressure as she attempts to complete a history assignment on the Gallic Wars and the collapse of the Western Roman Empire—a subject she would ordinarily find fascinating.

Don is reading an editorial about climate change in the Sunday newspaper, while Mary is reviewing an investment portfolio. After finishing the article, Don places the newspaper on the coffee table with a slight sigh of exasperation. He excuses himself, saying he's going to practice a new Celtic music composition that he's written expressly for the erhu.

Caroline senses an opportunity to break away from her assignment.

Caroline: "Dad, you don't look happy about the news."

Her father replies that it's the climate change controversy.

Don: "It's like a religious war."

Caroline: "I've heard climate change mentioned on television, but I'm not sure I know what it means. What's changing?"

Don: "That's a good question. For one thing, the Earth's average temperature is increasing. But…"

Caroline: "You mean … the Earth has a fever?"

A Small Increase in Average Global Temperature Can Radically Disrupt the Planetary Ecology

Don: "That's a reasonable, though distressing, analogy. Just as a viral or bacterial infection has multiple consequences—sore throats, headaches, muscle aches, and so on—a change in planetary temperature can have a wide range of effects, such as rising sea levels, shifting vegetation patterns, and more intense storms."

Caroline (interrupting): "To varying degrees? Is that a pun?"

Her father continues with a light smirk. "As I was saying, a small increase in average global temperature can radically disrupt the planetary ecology to which humans have adapted over the past 10,000 years."

Caroline: "Sorry, Dad, you're losing me. I don't know what planetary ecology means."

Don: "Ecology refers to how living organisms interact with one another and their environment. The largest and most complex of those organisms is the Earth itself."

Caroline: "The Earth is an organism?"

Don: "Yes, it's a living organism. I like to call it organ-earth. We're part of it, just like plants, microbes, the oceans, the atmosphere, other animals, and all living and non-living matter. What's more, they're all interconnected, continuously changing, and affecting each other—sometimes positively, sometimes negatively. So, it's a very complex and dynamic organism."

Caroline: "What do you mean when you say the Earth is a complex, dynamic organism?"

Don: "The meaning should be clear from what I just said, but let's back up a bit. In the 19th century, two mathematicians—Alfred James Lotka, an American, and Vito Volterra, an Italian—independently published mathematical equations describing the interdependence of predators and their prey. As the predator population increases, the prey population declines. But as the prey population declines, the predator's food source shrinks, and so the predator population starts to decline, allowing the prey population to recover.

"The equations illustrate what engineers call a two-variable nonlinear model with negative feedback, and they correctly predict stable cycles: each population increases and then decreases in a regular pattern."

Caroline: "Dad, I have no idea what you're talking about."

Don: "Take a guess. What do you think a two-variable nonlinear model with negative feedback means?"

Caroline (haltingly): "I guess the variables represent the two populations, but I'm confused by the rest of it."

Don: "Think about the possible meaning of feedback."

Caroline: "Oh, I think I've heard one of my teachers ask the class for feedback. Does it mean that when someone does something, another person responds to change the behavior of the first person?"

Don: "Yes, but it doesn't have to involve people. It could be animals or mechanical devices. For example, a toilet is an example of a negative feedback system—and you might want to Google it to find out why that's the case. Let's call the interacting pair agents and say that one agent responds to another in a way that changes the behavior of the first agent. If the change in behavior is amplified, that's positive feedback; if it's diminished, that's negative feedback."

Caroline: "I think I understand that, but what does nonlinear mean?"

Don: "Let's think about what you learned in algebra. I'm sure you've drawn graphs representing the equation $y = ax + b$, and you've also graphed equations like $y = ax^2 + bx + c$. The first is a linear equation because when you plot y $a$gainst x, you get a straight line. The second equation does not represent a straight line, so it's nonlinear. In this case, it's nonlinear because it depends on x^2."

Caroline (haltingly): "I think I get it. Does the variable representing the population of predators multiply the variable representing the population of prey?"

Don: "Yes, that's the origin of the nonlinearity. But I don't want to go into too much detail. I just want to emphasize that the behavior of systems of equations that are nonlinear and have negative feedback is difficult to intuit. For example, Volterra's interest was sparked when his son-in-law, a marine biologist, told him that during the First World War, the percentage of predator fish caught increased. That was puzzling because fishing diminished considerably during the war, and since prey were the preferred catch, biologists expected the prey population to increase relative to the predators. It was the mutual interdependence of the two populations that explained the observation."

Don continues: "Now here's the thing. The Lotka-Volterra equations represent a very simple two-variable model. If you imagine a network of many interacting populations with various types of feedback between them, it would be impossible to predict the behavior—and for reasons I won't go into, it would be difficult even with a powerful computer. The bottom line is that no one can fully grasp how something as complicated as organ-Earth will behave if it's perturbed.

"But let's not get sidetracked. For now, there are two takeaway messages: first, the Earth is a large, complex organism, and even relatively small, sustained perturbations can cause large destabilizing changes; and second, scientists have uncovered a lot of evidence indicating that human activities are perturbing the planetary ecology, although not everyone thinks the change is serious or that it's happening as rapidly as some people say it is."

Caroline: "What is the evidence—and why doesn't everyone believe that a changing climate is a serious problem?"

Don: "Tens of thousands of scientific studies carried out in thousands of laboratories overwhelmingly indicate that our climate is changing. But although the science is clear, controversy arises when people start thinking about the costs of doing something about it and weighing those costs against other needs. Mitigating climate change is very expensive, and anything we do won't have an effect for decades. That's why some people think we're better off spending the money solving problems that will have an effect in the immediate future—problems like oppression, infectious disease, education, and various crises that are affecting hundreds of millions of people right now. It's very complicated: the economics, the ethics, the politics."

Mary (who had been balancing a stock portfolio but was also paying attention, couldn't resist chiming in): "But Don, isn't it true that the cost of doing nothing could be even higher? Rising sea levels could wipe out entire cities."

Don: "You have a point—and, of course, that's part of the reason developing policy is so complex; it's a tangled web of economics, ethics, and politics."

Caroline: "I don't get it. How does that explain why some people are doubters? Economics, ethics, and politics can't change the facts."

Don: "It's true that political bias doesn't change the facts, but it does change how people react to the facts. In any case, discussing politics will take us too far off topic."

Caroline: "So what happens if we don't act? Does our way of life disappear? What's the cost of saving the planet compared to the cost of losing it?"

Don met her gaze: "That's what scientists are trying to figure out. The Earth is a complex system, and small changes can lead to big, unpredictable consequences. That's why we need to listen to the experts."

Mary: "Your aunt's fiancé, Barry Bruce, is a professor of geology, and he knows a lot more about climate science than I do. We're going to have dinner with him and Aunt Zoe on Friday, and like your aunt, he loves to interact with students. I'm sure he'd be happy to answer your questions."

Caroline leaned back, her history assignment forgotten. "I think I'm going to have a lot of questions for Barry on Friday."

Several Successive Years of Declining Temperature Does Not Mean the Fever is Subsiding

Caroline's week had been a typical whirlwind in the life of a vibrant teenager—playing soccer with her friends, practicing the violin, bombarding her teachers with question after question, and savoring the simple, surprising experiences of early adolescence that she could hardly have imagined six months ago. Now it's Friday evening and Caroline, her mother, her father, her aunt, and Barry are gathered around the Angstroms' dining room table, finishing dessert. Caroline is uncharacteristically quiet.

Don: "Caroline, do you still have questions about climate change? Now's your chance."

Caroline: "I guess so, but I had such an intense week at school that I forgot what I wanted to ask last Sunday."

Don: "You asked what it means when someone says that the atmosphere, land, and ocean interact."

Caroline: "Oh, right. I still have only a very vague idea about what that means—but I just remembered a simpler question I wanted to ask."

Barry (sympathetically): "What's the question?"

Caroline: "I get the impression that a lot of people, including congressmen, don't believe the planet's temperature is increasing. For example, they say that in some years, the temperature has been lower than in previous years. Doesn't that make you wonder—is the Earth's temperature really increasing?"

Barry (nodding thoughtfully): "It does make some people wonder if the climate is really changing. But no, it doesn't make me wonder."

Caroline: "Why not?"

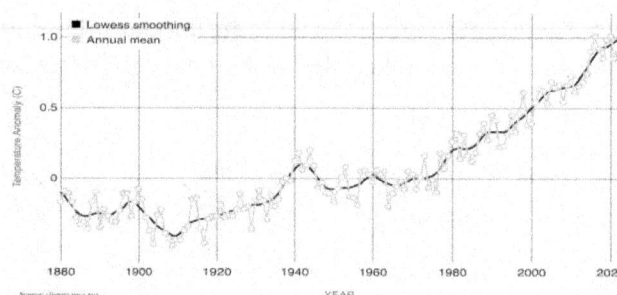

Figure 1. The global land and sea average temperature relative to the average temperature between 1951 and 1980.

Rather than responding immediately, Barry pulls out his phone and Googles data from the National Aeronautics and Space Administration (NASA). Within moments, he retrieves a graph of the average annual global temperatures from 1880 to 2021, relative to the average temperature between 1951 and 1980.

Caroline: "Why is there so much irregularity?"

Barry: "The interplay between the oceans, the atmosphere, and the land is very complicated. If you do some research, you'll find well-known oceanographic and atmospheric phenomena that create erratic changes in global temperature. However, superimposed upon these natural variations is a long-term trend in temperature, which becomes clear when you look at data over an extended period. This again illustrates the difference between weather and climate—to see climatic changes, you need to examine data spanning decades."

Natural Processes as Well as Human Activities Drive Long-Term Temperature Trends

Caroline: "Aren't weather and climate related?"

Barry: "Yes, they're related—although weather generally refers to changes occurring more or less randomly over short periods, such as days, weeks, or months. In contrast, climate involves changes in the entire planetary system—atmosphere, biosphere, oceans, rivers and lakes, the Earth's crust, and so on—which occur over decades, centuries, or even millennia. During the past 60 years, the Earth's average surface temperature has increased by about 0.7°C, and over the past 150 years or so, it has risen approximately 1.2°C."

Caroline: "I'm confused. I understand your explanation of the difference between weather and climate, and I don't doubt that the average global temperature has increased a bit during the last several decades. But how do you know that the change we're experiencing is the result of human activities rather than something that just occurs naturally? Can't natural changes produce long-term trends in climate? In other words, how can you be sure that this increase we're experiencing won't reverse itself in, say, 50 or 100 years? And what are these human activities that you're referring to?"

Natural Changes in Climate Are Easily Distinguished From Anthropogenically Driven Change

Barry: "Those are great questions. Your first question can be answered very simply: yes, natural geological changes do have long-term effects on climate. The Earth's climate has changed radically and continually in all sorts of ways throughout its history."

Barry: "It's startling to realize that around 4 billion years ago, when the Earth was formed, it was nothing but a barren, lifeless rock. Then, a short time later—perhaps after a few hundred million years—molecules started to form and replicate. Those profoundly mysterious events marked the beginning of life. Eventually, very simple replicating cells emerged, followed by multicellular organisms, animals with brains, and ultimately beings with the ability to understand. The Earth evolved from a lifeless rock into a complex living organism trying to understand and regulate itself. No one knows exactly how the process started—it's one of the great mysteries of science—but we do understand something about how it progressed and how our species, *Homo sapiens*, came to dominate the planet."

Barry (half-apologetically): "But I'm digressing a bit."

154

Caroline: "No, this is fascinating—please continue."

Barry: "The causes and characteristics of the deep and wide changes in the Earth's climate system, which have occurred for billions of years, are actually quite distinct from the changes caused by human activity—the so-called anthropogenically driven changes."

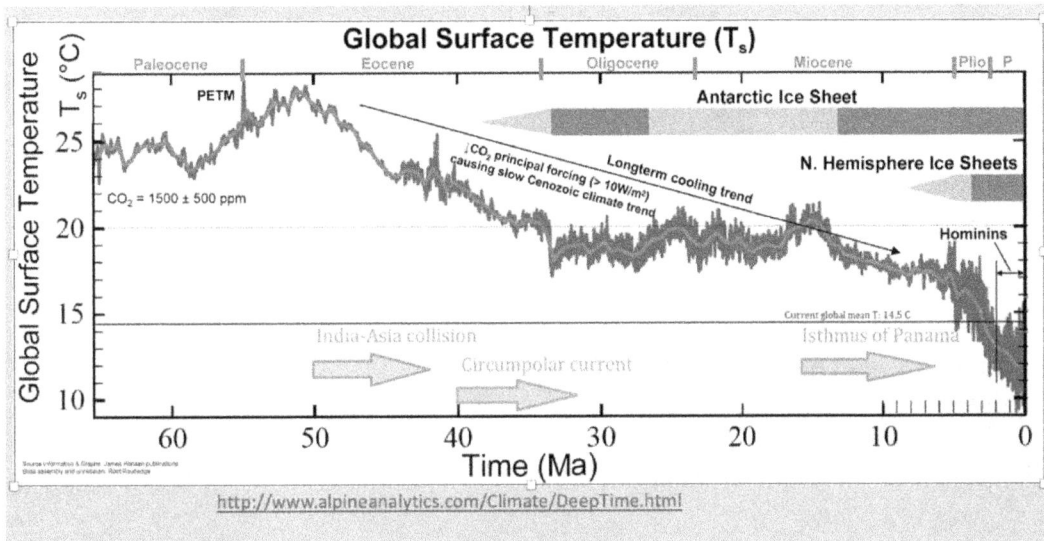

Figure 2. The planet has undergone climatic upheavals throughout its 4 Billion year history, with the profound changes in different geological periods generally occurring for different reasons. This figure summarizes the pattern of global temperature variation during the Cenozoic era, and some of the processes driving it.

Barry then searched the Internet for an example and found a graph showing the average global surface temperature during the Cenozoic era, which started 65 million years ago, around the time of the great dinosaur extinction. The graph illustrated large and erratic temperature changes— approximately 15°C—over the past 50 million years and indicated the drivers of those changes. For instance, the collision between the Indian subcontinent and the Eurasian plate about 40 million years ago formed the Himalayas and the Tibetan Plateau. Among the consequences of that collision were altered rain patterns and chemical weathering, which drew down atmospheric carbon and drove a long-term decline in temperature.

Caroline: "I'm lost—Eurasian plate? Chemical weathering? I have almost no idea what they refer to, although I've heard some of my classmates who are taking Earth Science use the words when they talk to one another."

Barry: "Oh, I didn't realize you hadn't taken Earth Science yet. When you do, you'll learn that one of the most important geological theories of the past 150 years is that Earth's continents move, occasionally colliding and reshaping the planet's surface. This groundbreaking idea has its origins in a proposal made roughly a century ago by Alfred Wegener, who hypothesized that continents drift slowly across the Earth's surface, much like plates floating on a liquid layer. At the time, this notion was considered highly radical and faced widespread rejection, as it implied that the Earth's lithosphere—the rigid outermost layer—doesn't behave as a solid, unchanging structure, but as a system capable of slow, fluid-like motion over geological timescales. However, in the decades that followed, a wealth of evidence from various measurements and discoveries confirmed the theory, which is now known as plate tectonics. It stimulated research that revealed a lithosphere of segregated rigid tectonic plates that do in fact move, albeit at a very gradual pace."

Caroline: "OK, great, that's very clear. And chemical weathering?"

Barry: "You can probably take a reasonable guess at its meaning: it's just the breakdown of rocks and minerals by chemical reactions that alter their composition and create new minerals over geological time periods. In essence, it's how the Earth's surface gets slowly reshaped by chemical processes."

Caroline: "Got it!"

Barry then continued with his description of changes in the planetary surface over the past tens of millions of years.

"Another important event was the formation of the Isthmus of Panama, which had a major impact on global climate and connected two continents. This led to, among other things, the creation of the Gulf Stream and large-scale animal migrations.

"This is just one of many periods of change in the Earth's history. None of these changes, though, can be used to rebut the idea that the current climate change is natural. Even the spike denoted by the Paleocene-Eocene Thermal Maximum (PETM), which lasted 170,000 years and during which global temperatures rose by somewhere between 5 and 9°C, doesn't have the same characteristics as the current rate of change."

Barry was enjoying the exchanges, but also tiring—which was quite remarkable considering he was a man of considerable energy. He was, however, ready to say goodnight and as he did so he said he would send Caroline some reading material, feeling confident that she'd find the story of our planet's geology and ecology fascinating.

Greenhouse Gases Absorb Heat

Zoe, who had been following the conversation closely, and didn't seem ready to call it a night, jumped in spontaneously.

Zoe: "Caroline, have you heard of greenhouse gases?"

Caroline: "I've heard the term, but all I know is that people say they have something to do with climate change."

Barry sat down again and volunteered that "Greenhouse gases are small, heat-absorbing molecules, such as carbon dioxide (CO_2), methane (CH_4), and nitrous oxide (NO_2). They're present in the atmosphere at extremely low concentrations—what we call trace quantities. The atmospheric concentration of carbon dioxide, the most abundant of these gases, is 420 parts per million (ppm), while the concentrations of methane and nitrous oxide are even lower—so low that they're measured in parts per billion (ppb). The relevance of greenhouse gases to the rising temperatures we're experiencing is that they absorb heat and..."

Caroline (interrupting): "I'm confused already. Parts per million? Parts per billion? Of what?"

Barry: "Good catch. 420 parts per million means that for every 1 million air molecules, 420 of them are carbon dioxide molecules."

Don (looking at his daughter): "Can you convert that into a percentage?"

Caroline: "Dad, that's a third-grade problem. 420 ppm means $420/1,000,000 = 0.000420 = 0.042\%$."

Zoe: "Getting back to greenhouse gases—have you studied the Industrial Revolution in your social studies class?"

Caroline: "We're just about to start. But what does that have to do with greenhouse gases?"

Barry, now resigned to stay a bit longer: "A lot, as we'll explain in a minute."

Zoe: "The Industrial Revolution began when people invented methods to convert heat energy into mechanical energy inexpensively."

Caroline: "Heat energy? Mechanical energy? I'm lost."

Mary, trying to recall her undergraduate thermodynamics course in aerospace engineering: "I know the feeling. Let me try to explain. If you heat a gas-filled cylinder with a movable piston, the pressure created by the expanding gas will move the piston: heat is converted into mechanical energy—the motion of the piston. When several pistons work together, the mechanical energy they generate can do things like turn the wheels of a train or paddle a boat.

"In other words, heat is converted to mechanical energy that can do work. The ability to convert heat efficiently and economically enabled the development of machines that carried out tasks previously performed by humans or animals. This revolutionized industries like transportation and textiles, leading to massive economic growth and increased use of fossil fuels, which, as we'll see, contributed to rising levels of greenhouse gases."

Caroline: "OK, I get it. But where do greenhouse gases come in—and what does all this have to do with climate trends?"

Barry (continuing with some history): "All the new technologies developed during the Industrial Revolution required energy, which was produced by burning fossil fuels."

Caroline: "Fossil fuels? What are they?"

Zoe: "Plants and animals contain large amounts of carbon."

Caroline (surprised): "They do?"

Zoe: "Yes. I'm sure you've learned about DNA and proteins."

Caroline: "A little."

Zoe: "Well, they're made mostly of carbon, oxygen, nitrogen, and hydrogen. Coal, for example, is the remains of plants and animals that decayed and were compressed for millions of years deep beneath the Earth's surface. When coal is burned, the carbon combines with oxygen to form carbon dioxide (CO_2). While coal is a plentiful and inexpensive source of heat, the amount burned to generate power for manufacturing, construction, heating, and other activities has increased dramatically since the start of the Industrial Revolution about 170 years ago. As a result, the rate at which greenhouse gases are entering the atmosphere is also rising. And there's another downside: it doesn't burn completely, and the byproducts of incomplete combustion are toxic."

Caroline: "Are you saying that burning coal not only produces CO_2—which is necessary for animal and plant life—but also emits pollutants into the atmosphere?"

Zoe: "Yes, burning fossil fuels emits pollutants. If I remember correctly, the World Health Organization estimated that in 2019, more than 4 million deaths were attributable to outdoor air pollution—and that might be an underestimate. A 2021 study by researchers at Harvard, Birmingham, Leicester, and University College London concluded that in 2018, 8 million deaths were caused by toxic byproducts of fossil fuel combustion. But let's not digress—that's another story entirely."

Barry: "Keep in mind, the health hazards are in addition to the global warming caused by increasing atmospheric concentrations of heat-absorbing greenhouse gases."

Barry continues: "By the late 1800s, coal was being used to generate electricity for factories that revolutionized the manufacturing of textiles (fabrics). It also became a common energy source for rail travel and home heating. As countries in Western Europe and the United States industrialized, the rate at which carbon entered the atmosphere increased rapidly."

Caroline: "I have a question. I just learned in social studies that there are more than 8 billion people on the planet—and I know from science class that we exhale carbon dioxide. Wouldn't that be a significant contributor to the increase in atmospheric CO_2?"

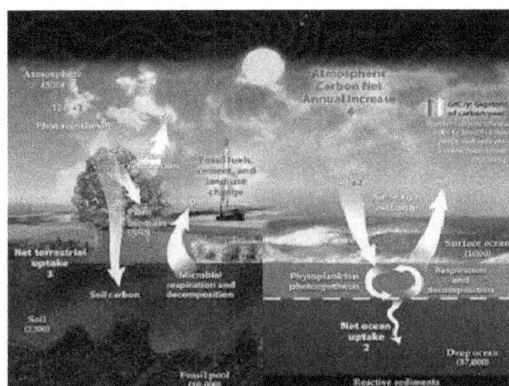

Figure 3. A simplified summary of carbon exchange between four of its reservoirs: land, atmosphere, ocean and deep ocean. Although the figure represents the dynamic exchange faithfully, the numbers are out of date. Significantly, the 9 gigatons of carbon released into the atmosphere by human use is now closer to 11 gigatons, and the total atmospheric concentration is approximately 870 gigatons. Of greater importance is that the fraction of carbon retained by the atmosphere remains at approximately half of what is released, with three of the 9 gigatons being returned to the soil and two being returned to the ocean.

Barry is surprised by the question, which indicates Caroline might not fully understand what she's learned in school. Her aunt seems surprised too.

Zoe: "You've learned about the carbon cycle, haven't you?"

Caroline: "Yes, we're learning about it now."

Zoe: "Then you know that carbon cycles between the air, plants, and animals. Plants convert the inorganic carbon in atmospheric CO_2 into the carbon present in organic molecules like proteins, nucleic acids, and carbohydrates. Animals then eat the plants, digest them, and convert the organic carbon back into CO_2, which they exhale."

Caroline: "Oh, now I get it! Animals and plants don't create carbon—it just passes through them, first as carbon dioxide, then as part of biological molecules, and then back to carbon dioxide in the atmosphere. Is that what you meant by the interaction between the atmosphere, plants, and animals?"
Barry and Don: "Yes, that's part of it."

Infrared Radiation Emitted by the Earth is Absorbed and Reemitted Many Times by Greenhouse Gases Before Escaping the Planet

Caroline: "But I still don't understand how an increase in greenhouse gases leads to global warming."

Barry: "I know, I was about to explain that before we got sidetracked."

Don (turning to his daughter): "Do you remember the conversation we had about light—infrared, ultraviolet, visible, and so on?"

Caroline: "Yes."

Don: "Well, the thing about CO_2 and other greenhouse gases is that they absorb infrared radiation, and most of the heat radiated by the Earth's surface is in the infrared frequency range. As infrared radiation travels toward space, it's absorbed and reemitted by CO_2 many times—in all directions—before finally escaping the atmosphere."

Caroline: "So it takes a lot longer for heat to leave the planet than it would if there were no greenhouse gases?"

In the Absence of Greenhouse Gases, the Average Surface Temperature of the Earth Would be ~ -18°C

Barry: "Yes. Without greenhouse gases, energy would arrive at and leave the planet at the same rate, assuming the Sun's energy remained constant. That would also be true with a constant amount of greenhouse gases; in both cases, the planet would be in thermal equilibrium. However, without greenhouse gases, the planet would be much colder."

Caroline: "Thermal equilibrium? What does that mean?"

Barry: "Sometimes you can infer meaning from context. What do you think it means?"

Caroline: "The rate at which energy leaves the planet is the same as the rate at which it arrives—I guess."

Don: "Yes, that's what it means. You shouldn't have to add 'I guess.'"

Caroline: "OK, I guess you're right."

Don: "Touché."

Caroline (with a puzzled look): "But the concentrations of these gases are so low. I'm surprised they can have much of an effect. How much colder would the planet be without them?"

Barry: "I'm sure I came across that somewhere, but I don't remember the exact number."

Caroline: "Would it be difficult to calculate?"

Barry: "I think the physics would be a bit challenging for someone at your grade level, but if you'd like, we can give it a try."

Caroline: "Great! I'd like to take the plunge."

Barry: "Alright, let's start by defining the Earth system as all its interacting components—atmosphere, water, living organisms, land, and ice. But in this case, there's no atmosphere. Among the uncommon words that are used in a calculation of this sort is energy flux: the energy per unit time per unit area. In other words, it's the power per unit area. I'll use 'power,' 'energy current,' or just 'current' interchangeably."

Caroline: "When I think of current, I always think of electricity."

Barry: "Current is a general term that describes the rate at which anything flows. In electronics, it's electrons per second; in chemistry, it could be the number of molecules per second. In our case, it's the flow of energy, or joules per second. The quantity of interest is the (energy) flux falling on each square meter of the Earth system, but to calculate that, we begin by determining the flux falling on the Earth's orbital sphere.

"Essentially, this is a sphere with a radius equal to the distance from the Sun to the Earth, centered on the Sun. The flux falling on this orbital sphere is called the solar constant, which we'll denote as S. Given this definition, and letting r denote the radius of the Earth, the power (or energy current) entering the Earth system, if none were reflected back into space, is $S\pi r^2$—this represents the current intercepted by the Earth.

"Of course, a fraction of the power, which we'll denote as α, is reflected back into space. That means the current entering the Earth system is $S\pi r^2 (1 - \alpha)$, and the quantity we're looking for, the flux (E_s) entering the Earth system, is the current divided by the area of the Earth:

$$E_s \sim S(1 - \alpha) / 4.$$

"If you look up the values of S and α, you'll find that $E_s \sim 238$ W/m², where $\sim$ means 'approximately.'

"I know this can be a little confusing, and it's about to get even more abstract—should I continue?"

Caroline: "I need to think about what you just said, but I think it's best if you continue now so I can see the entire picture all at once."

Barry: "OK, let's proceed. To calculate the Earth's temperature in the absence of an atmosphere, we need to know not only the flux entering the Earth system but also the flux leaving it. Now we're diving into some new physics—physics developed by Max Planck. You know who he was, right?"

Caroline is just about to reply when Barry continues.

"Here's what's important for our conversation: the spectrum—energy as a function of frequency— of the Earth's emitted energy can be approximated by that of a blackbody."

Barry pauses thoughtfully, and then continues:

"I know what you're going to ask—you want to know what a blackbody is. The answer is simple: it's an idealized object that absorbs all incident electromagnetic energy and reemits it as a continuous of electromagnetic radiation that depends solely on the temperature of the object. Planck developed the mathematical theory of this spectrum based on the quantum hypothesis, solving a longstanding puzzle commonly referred to as the 'ultraviolet catastrophe.' Most deviations from a perfect blackbody can be accounted for by introducing a quantity called the effective emissivity. The Earth's effective emissivity, which is slightly less than one, accounts for atmospheric influences, such as greenhouse gases and clouds, which absorb and reemit radiation."

Caroline: "Ultraviolet catastrophe? That sounds intriguing—what is it?"

Barry: "I'll leave that for you to find out. When you do, you'll learn about one of the most exciting and revolutionary periods in physics—and maybe even pick up a little quantum physics along the way. For now, though, I'll tell you that an important relationship between energy flux and temperature can be derived from Planck's theory. It's called the Stefan-Boltzmann equation, and it states that the energy flux from a blackbody is proportional to the fourth power of its temperature. Consequently, in the absence of an atmosphere, when we equate the two fluxes, we have:

$$\frac{1 - \alpha}{4} = \varepsilon \sigma T^4$$

"where σ is the Stefan-Boltzmann constant, and ε is the emissivity. I'm sure you know how to solve this equation for the temperature, so go ahead and give it a try. When you're finished, let me know the answer."

Caroline leaves the table and returns 10 minutes later with an answer.

Caroline: "In the absence of an atmosphere, and taking $\varepsilon = 1$, the Earth's temperature would be 18° below 0 on the centigrade scale (-18°C), which is about 32°C colder than the Earth's current temperature."

Don (smiling with a half-jocular tone): "And that doesn't even include the wind-chill factor."

Caroline (nodding): "I guess human civilization would have had a very difficult time developing without greenhouse gases."

Zoe: "And not just because the entire planet would be frozen—the entire ecosystem, including how millions of species develop and interact, would have been radically different."

Caroline: "We've gone off on a lot of tangents..."

Don: "Yes, because you can't contain your questions—but that's OK."

Caroline: "I just want to make sure I understand why the long-term trend is really a trend. I think what you're saying is that since the temperature increase is due to the rising levels of greenhouse gases—mostly CO_2—and the CO_2 increase is a result of the rapid rise in the use of fossil fuels driven by the Industrial Revolution, the planet's temperature will continue to rise as long as countries continue to industrialize using coal and other fossil fuels."

Barry: "Yes, that's pretty much the idea. As things currently stand, the temperature will continue to increase unless we either stop greenhouse gas emissions, find ways to remove greenhouse gases from the atmosphere, or find ways to cool the planet—or maybe all three."

Caroline: "Find ways to remove greenhouse gases from the atmosphere? Is that even possible?"

Barry: "Yes, it is. Removal is a crucial part of any effective strategy for reducing the concentration of atmospheric greenhouse gases, but let's discuss that in more detail later."

Caroline: "OK, I believe that the increase in CO_2 contributes to the rise in temperature, but is it enough to explain the entire temperature increase? And how quickly is CO_2 increasing?"

Barry: "Let's start with the second question first, because it's easier to answer. Records of atmospheric CO_2 date back at least to the mid-18th century. While the number of observations and their accuracy weren't as reliable back then as they are now, the pre-industrial level is generally accepted to have been around 278±2 ppm. With that baseline, we can now compare temperature changes to changes in CO_2 concentration over the past six or seven decades.

"In the late 1950s, a geophysicist named Charles Keeling began measuring atmospheric CO_2 levels atop Mauna Loa in Hawaii using modern instruments."

Barry pulls out his cell phone and shows Caroline the famous Keeling Curve—a graph depicting annual atmospheric CO_2 concentrations in Hawaii from 1959 to 2020.

Caroline: "Why do the lines look jagged? Is that because of errors in the data?"

Barry: "No, the ups and downs are too regular to be errors. The most likely reason is that plants absorb atmospheric CO_2 during their growing season and then release it back into the atmosphere during the fall and winter."

Caroline: "The precision of the data is impressive. Did you know Charles Keeling? He was a geophysicist, just like you, right?"

Barry: "Yes, I met Charles Keeling once, and it was a life-altering experience.

"I grew up in a low-income, African American community in East Harlem, which had a rich and vibrant culture. We all knew our neighbors, and we felt deeply connected. One of the things that influenced me most growing up was music, especially jazz. Its complex key changes, varied rhythms, and spontaneous improvisations had a huge impact on me. But beyond that, I was also very good at mathematics, and even back in the early '90s, climate change was starting to grab headlines. I wanted to understand it better, so I studied mathematics and the sciences."

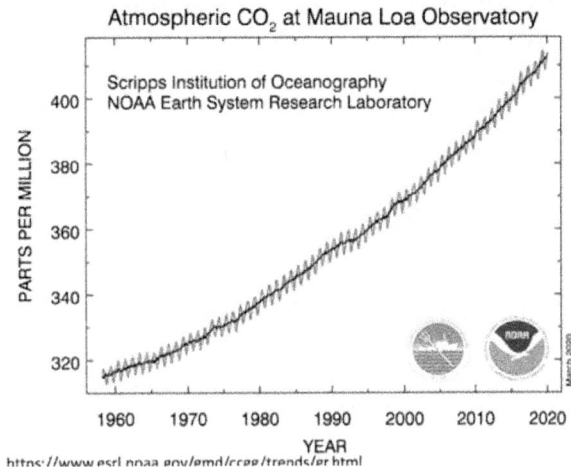

Figure 4. A Keeling curve showing a regular and dramatic increase in carbon dioxide in the northern hemisphere over a period of six decades.

Barry: "I went to The City College of New York—which is in Harlem and has many distinguished alumni, including 11 Nobel Laureates. I majored in music with a minor in geology but when I graduated I didn't know what I wanted to do. Although I loved music, I didn't think I was talented enough to have a successful career, so I wound up applying to PhD programs in geophysics. One of those programs was at the Scripps Institution of Oceanography. Charles Keeling was a professor there, and when he saw my application, he invited me to visit."

Caroline: "Wow!"

Barry: "Yes, that visit changed my life. Before I met him, I wasn't really sure I wanted to become a scientist, but after meeting him, I had no doubt that I wanted to follow in his footsteps.

"Keeling died in 2005, but the measurements continued, and now observatories all over the world collect greenhouse gas data. In 1958, when the study began, the atmospheric concentration of CO_2 was 315 ppm. By 2017, the concentration had risen to 405 ppm, and in 2023, it hit a record high of 420 ppm."

An expression of shock bordering on fear crossed Caroline's face as she asked almost timidly, hoping to be contradicted: "Isn't that incredibly fast—is it accelerating?"

Taking out her iPhone, Caroline adds: "That means it has increased by about 33% in the last 65 years."

Barry: "How much did it increase in the 80 years before that?"

Caroline: "Let me see... (315−278)/278×100 = 13%."

Caroline is in a state of disbelief as she confronts the fact that the increase is not just accelerating but accelerating rapidly. Half-hoping there's a mistake somewhere, she asks with uncharacteristic timidity how scientists determine CO_2 levels.

Caroline: "How do scientists determine CO_2 levels?"

Barry: "Think! What do you know about CO_2?"

Caroline: "That it's a gas."

Barry: "Yes, but what do you know about its properties?"

Caroline: "It's colorless, odorless, and absorbs heat?"

Barry: "Yes, it absorbs heat in the form of... what kind of radiation?"

Caroline: "Infrared."

Barry: "Exactly. And if infrared light is passed through vessels containing various amounts of CO_2, the concentrations can be calculated based on the differences in intensity between the incoming and outgoing beams. I'll leave the details for you to research later."

The Atmospheric Concentration of CO_2 Cycled Between 125 ppm and 290 ppm for 800 Thousand Years Then Spiked to 420 ppm During the Last 150 Years

Caroline: "Just one question."

Barry: "Of course."

Caroline: "Is the same method used for all greenhouse gases? And if the vessel contains air, there are other greenhouse gases present too. How can you tell which gases are absorbing heat and how much?"

Don: "We'll leave that one for you to figure out."

Caroline: "OK, I'll think about it—but since we've covered so many topics, I'd like to make sure I understand the main ideas."

Barry: "Sure, go ahead and summarize what you've learned."

Caroline: "First, you told me that the atmospheric concentration of heat-absorbing gases has increased significantly over the past 150 years. Second, because of this, the rate at which heat leaves the planet is slower than the rate at which it arrives. Third, this energy imbalance is driving the observed increase in global average temperatures. Finally, the rising temperature is having a profound impact on the planet's ecology."

Mary: "Great summary! I think you've got it."

Caroline: "One thing still seems strange: 1.2°C doesn't sound like much of a temperature difference. It's much smaller than the difference between daytime and nighttime temperatures, and way smaller than the temperature difference between different parts of the world."

Barry: "Remember, there's a difference between weather and climate: 1.2°C is a global average temperature change spread over many decades, and it's layered on top of the natural fluctuations in temperature that happen daily, monthly, and annually."

Caroline: "OK, now I see. The difference between day and night and between temperatures in different regions continues as usual, but on top of that, there's an overall increase in temperature. It's like a rising tide."

Barry: "Yes, that's a reasonable way to put it. Let me remind you of two numbers that will help put the scale of the energy imbalance into perspective. The first number is solar irradiance—the energy flux arriving at Earth from the Sun. We already derived an equation for that, and it gives us a value of 238 W/m². The second number is the difference between solar irradiance and the energy transmitted back into space from the Earth—in other words, the energy imbalance. Our best estimate of the imbalance is about 0.9 W/m²."

Caroline: "The energy imbalance is considerably less than 1% of the solar irradiance—that seems really small."

Barry: "It doesn't seem like much because the numbers I gave you refer to power per square meter of the Earth's surface. But when you multiply that by the Earth's surface area—5×10^{14} m²—you'll see that the heat being added to the planetary system each second is approximately equal to the power output of four or five atomic bombs dropped on Hiroshima."

Caroline is shocked, especially when she takes out her calculator and finds that the amount of heat added to the Earth system each day exceeds the heat produced by more than 86,000 atomic bombs. She now asks the opposite question.

Caroline: "Why haven't we been burned to a crisp?"

Barry: "The answer is complicated, but basically it's because the oceans absorb about 90% of the heat, and the energy is spread out over the entire surface area of the planet. The warming of the ocean, though slow, opens up another whole set of problems, but for now, let's set those aside and return to the relationship between greenhouse gases and the Industrial Revolution. I'll keep it simple by just talking about CO_2."

Caroline: "There are still a couple of things that trouble me. First, I'd like to understand the relationship between the energy imbalance and temperature change. Second, I'm still not fully convinced that the current changes can't be driven by natural processes. You gave me a really interesting history of large-scale geological processes driving profound changes in temperature and geography over the past 65 million years, but that doesn't necessarily mean there aren't other natural processes that could explain the current changes."

Barry: "You're absolutely right to consider that there might be other changes similar to what's happening now.

"Let's begin by looking at relatively recent history—at least from the perspective of paleontology and geology. Over the past 800,000 years, Earth's temperature and atmospheric greenhouse gas concentrations have cycled in sync, with a dominant cycle of about 100,000 years. The temperature fluctuated between approximately 2°C above and 7°C below its 13.7°C average value in 1850, just before the start of the Industrial Revolution.

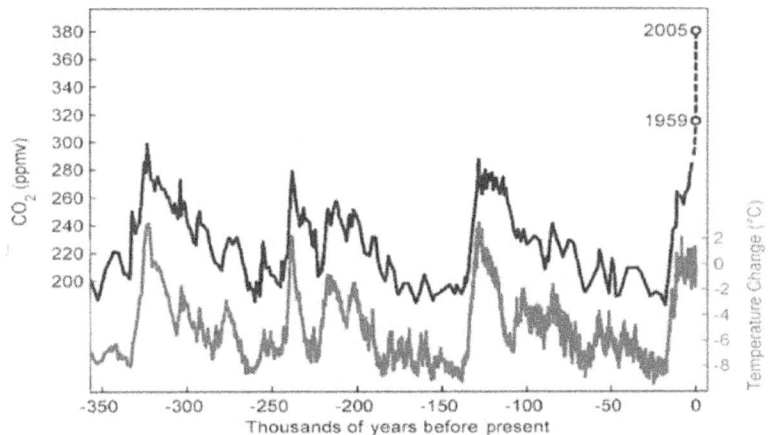

Source: J. R. Petit and others, "Climate and Atmospheric History of the Past 420,000 Years from the Vostok Ice Core, Antarctica," *Nature* 399 (June 1999): 429–36.

Figure 5. Climatic cycles during the last 350 thousand years

"Here's a figure showing the last 350,000 years, which is pretty typical. The lower curve shows the temperature swings, while the upper curve tracks the atmospheric CO_2 concentration, which is cycling between 125 ppm and 290 ppm. The troughs of the temperature cycle correspond to glacial periods. The last one began about 110,000 years ago, reached its peak—called the last glacial maximum—around 22,000 years ago, and ended roughly 12,000 years ago. So, we're currently in an interglacial period, and no one knows how long it'll last. If the glacial period hadn't ended, civilization as we know it wouldn't exist."

Caroline: "What do you mean by a 'dominant period'?"

Don prompts his daughter: "Think about a musical note, like middle C on our piano, for example."

Caroline: "Oh, I think I get it. Middle C isn't just a single frequency; it has overtones, too."

Barry: "Exactly. In the natural world, light, sound—and even planetary orbits, if we consider long time scales—aren't strictly periodic. The cyclic behavior of glacial periods, in particular, is influenced by many factors. So when we say '100,000 years,' we're really referring to the dominant period within a much more complex set of cycles. What's especially significant is that after 800,000 years of CO_2 concentrations bouncing between 125 and 290 parts per million, it rose to 320 parts per million in just 150 years."

Caroline takes out her calculator and finds that this spike occurred during a period that's just one five-thousandth of the 800,000-year time frame. She's absolutely stunned by how abrupt it is.

Barry: "It's also worth noting that in the last 2,000 years CO_2 levels never increased by more than 30 ppm in any 1,000-year period."

Caroline: "That's fascinating. But if I can digress a bit, does anyone know what causes the 100,000-year cycle?"

Barry: "Not for certain, but it likely has to do with variations in Earth's orbital parameters."

Caroline: "Orbital parameters? What are they?"

Barry: "That's a subject in college-level mechanics, and if we dive into it, we'll never get anywhere else. Let me give you one example, though—and even with this one, I won't go into much detail. Do you know what an ellipse is?"

Caroline: "A stretched-out circle."

Don: "That's good enough for now. The Earth's orbit is elliptical, but its stretch—what mathematicians call its eccentricity—changes gradually and in a complex way over hundreds of thousands of years. In addition the tilt of the Earth's axis and its precession also change over long periods of time. A Serbian physicist named Milutin Milankovitch proposed that as these orbital parameters change, the average energy reaching Earth's surface changes, which leads to temperature changes and, consequently, to long-term climatic shifts—glacial and interglacial periods."

Caroline: "What about carbon dioxide? Couldn't changes in CO_2 have been responsible for the climate cycles?"

Barry: "That's a great question. CO_2 and other greenhouse gases did cycle during these glacial-interglacial periods, but they actually lagged temperature change, instead of leading it. And if you were to argue that greenhouse gases drove these long-term cycles, you'd still need to explain why they cycled in such a regular pattern over hundreds of thousands of years."

Caroline: "But now scientists say that carbon dioxide is driving the temperature change. It's confusing."

Barry: "Yes, and that difference is actually another indication that the current climate change and the 100,000-year cycles have different drivers."

Caroline: "Are the warmer interglacial periods like the one we're in longer than the glacial periods?"

Barry: "No, actually, most of the time large portions of the planet are covered by glaciers and ice sheets. About 22,000 years ago, during the last glacial period, ice covered all of Canada, and in some areas, it was more than 2 miles thick."

Caroline: "Ice sheet??"

Barry: "For now, just think of them as extremely large glaciers."

Caroline: "Can you be more quantitative?"

Barry: "Sure, How does 20,000 sq miles sound?"

Caroline's awestruck once again: "An area of 20,000 square miles and a thickness of 2 miles? Incredible! Did the ice reach New York?"

Barry: "Yes, during the last glacial maximum, New York City was covered by a sheet of ice around 2,000 feet thick—give or take 1,000 feet. In other words, it was likely thick enough to cover the entire Empire State Building."

Caroline is speechless for a moment and then offers an anticlimactic response:

Caroline: "So, scientists don't know exactly how deep the ice was?"

Barry: "That's right. We don't know the exact numbers. I should've emphasized that all these figures—the thickness of the ice, high and low temperatures, and so on—are approximate. For example, during glacial periods the temperature could have reached 6 ±2°C below current levels. Even though these numbers are approximate, they're still dramatic, and they clearly illustrate the scale of the changes."

Caroline: "The changes are shocking—do scientists think this cycling will continue?"

Barry: "Based on our current understanding of the underlying processes, it likely will. But no one can say for certain. We only know that over the past 800,000 years or so, the time between the minimum temperature of a glacial period and the maximum temperature of an interglacial period has typically been between 15,000 and 25,000 years. The temperature rises very quickly, but then drops slowly, with a lot of fluctuations along the way. Since the last minimum occurred around 24,000 years ago, a gradual decrease in Earth's temperature would actually be less surprising than an increase."

Atmospheric Conditions that Existed Millions of Years Ago Can be Inferred by Analyzing Air Bubbles Trapped in Ice

Caroline: "I'm curious about something else." And then, without waiting, she asks, "How can scientists possibly know what the temperature and CO_2 levels were 100,000 years ago?"

Barry: "That's a great question, Caroline. We can research that together on another occasion if you'd like to wait. Or, if you'd prefer to try finding out on your own, you can start by researching keywords like ice core, sediment core, and tree rings."

Caroline (with a pleading voice): "Come on, Barry, you're frustrating me!"

Barry: "OK, I'll give you some hints."

Barry continues: "Data on past climates can be obtained by a number of different methods. None of them are perfect, but all the data point to the same conclusions."

Caroline (feeling exuberant once again): "OK, give me an example!"

Barry: "A lot of information is obtained by removing cylinders of ice—so-called ice cores—from polar ice sheets."

Caroline (interrupting): "Ice sheets? Giant glaciers? Can you say more?"

Barry: "Ice sheets form when summer temperatures aren't high enough to melt the annual snowfall. When that happens, identifiable layers of ice form over the course of thousands of years, under the increasing weight of fallen snow. Temperatures, as well as the concentrations of various molecules in the atmosphere—carbon dioxide, nitrogen, water vapor, and so on—have been inferred from the contents of air bubbles trapped in layers of ice at different depths, as far back as 800,000 years. The procedure has been tested by analyzing bubbles formed several decades ago and comparing the composition of the air they contain to the known composition of air determined directly at the time the bubbles were formed."

At this point Barry's starting to fade and he tells Caroline that he'll leave it to her to research the methods for obtaining past temperatures from ice core data.

Caroline (so wrapped up in her thoughts that Barry's hint about continuing the conversation later goes unnoticed) turns to her parents.

"OK, let's say that current CO_2 levels are anomalously high and increasing, and some of the increase is due to human activities. How do you know that human activity explains most of the increase? You just said there are large natural cycles in atmospheric CO_2 concentration and temperature. They have nothing to do with human activity because there were no humans, let alone industries that emit CO_2."

Multiple Lines of Evidence Indicate That Modern Climate Change is Anthropogenically Driven

This time, Don—who understands all too well Barry's scarcely hidden fatigue—responds with the patience of a loving father: "Yes, my endlessly curious daughter, you're correct philosophically. There have been many times in the history of science when people were sure they understood something, but the correct explanation was beyond anything they could have imagined at the time. To say that the simultaneous increase in CO_2 and industrialization is unlikely to be a coincidence is not a rigorous scientific proof. And I suppose, if we want to be really skeptical, we can act unsurprised that after 800,000 years of stability a breakout in CO_2 and other greenhouse gases such as methane coincided by chance with the start of a technological revolution that required coal burning at unprecedented rates. But additional non-historical evidence is also available—and next week, if we have some time, I'll explain some of it, especially isotopic ratio data which is virtually unchallengeable. Right now, however, everyone is getting tired."

Caroline: "Isotopic ratios??"

Don: "Not now, Caroline, it's late, your aunt and Barry are both tired, and so are your mother and I."

Zoe and Barry thank Mary and Don for the dinner, and Don thanks Barry for explaining climate change to his daughter. Barry promises additional conversations in the future.

The week goes by rapidly. It's Saturday afternoon, and Caroline approaches her father, who's been practicing for a concert.

Caroline: "Dad? You said you'd tell me about isotopes?"

Don: "Sure, I can use a break anyway. Have you learned anything about isotopes?"

Caroline: "A little. I know that isotopes of an element have the same number of protons and electrons, but a different number of neutrons. We learned in my science class that deuterium, which is an isotope of hydrogen, has a nucleus with one proton and one neutron. Another hydrogen isotope, tritium, has one proton and two neutrons."

Don: "Great! Here's an example of additional evidence that fossil fuels are responsible for the increase in atmospheric carbon."

Don: "Carbon has several isotopes. The most abundant, carbon-12, has 6 protons and 6 neutrons. The carbon atom in most carbon dioxide molecules is carbon-12. However, some CO_2 molecules have a different isotope, carbon-13, which has 6 protons and 7 neutrons. Plants preferentially use the lighter isotope, carbon-12.

173

"The key observation is that the combustion of fossil fuels, which are plant products, produces CO_2 gas for which the ratio of carbon-13 to carbon-12 is lower than it is in the atmosphere. Over the past 70 years, as CO_2 has increased, the percentage of atmospheric CO_2 molecules containing carbon-13 is significantly lower than it was before the Industrial Revolution. This implies that the increase is mainly due to plant-derived carbon, i.e., fossil fuels."

Caroline: "So you're saying the evidence is beyond any reasonable doubt that CO_2 concentration is changing rapidly, and that the change is due to human activities?"

Don: "Yes. In addition, the rate at which atmospheric CO_2 is increasing tracks the rate at which CO_2 emissions are increasing."

Caroline: "I thought there was a lot of controversy. Are you saying there's little room for controversy?"

Don: "Controversy tends to collapse under the weight of overwhelming data, and the evidence linking the preponderance of CO_2 increase to human activity is very convincing. It's essentially universally accepted in the scientific community. You can always say there might be another explanation involving processes we don't fully understand, but why would you say that when we have a well-supported explanation? And if there is some mysterious phenomenon at work, it would, by definition, be non-actionable. Fortunately, we have an actionable explanation. Unfortunately, we're not acting on it."

Feedback Has a Pronounced Effect on the Rate at Which the Planetary Temperature is Increasing

Caroline: "That was an interesting digression into the earth's history, which provided helpful context for understanding the changing global temperature. I also understand how the difference between the incoming and outgoing energy flux will increase planetary temperature. But I still don't know how much the temperature will increase for a given difference in flux."

Don: "That's complicated, but I'll give you an approximate idea. Let's start with the temperature increase that's directly attributed to atmospheric CO_2. The physics of that process is well understood: doubling the CO2 concentration will increase the global average temperature by approximately 1°C."

Caroline: "An increase of one degree Celsius for each doubling doesn't sound like much. The increase has already been more than 1°C since the start of the Industrial Revolution, and the carbon dioxide level is less than double what it was."

Don: "You're right, but that 1°C increase is not the whole story—it's only a direct effect of the increase. It neglects feedback."

Caroline: "What do you mean by direct effect and feedback?"

Don: "What do you think I mean when I say that the 1°C increase is a direct effect of the increase in carbon dioxide?"

Caroline: "Does it mean that carbon dioxide increases the temperature of the surrounding air by reacting to infrared radiation?"

Don: "Yes, the CO_2 molecules absorb infrared radiation, causing the chemical bonds that hold carbon and oxygen together to vibrate more rapidly. They then release energy as heat as they relax back to their original state. Feedback means that the heat produced starts other processes in motion that generate more heat.

"One example is water vapor feedback. The maximum amount of water vapor that can be retained in the atmosphere is determined by temperature: the higher the temperature, the greater the water vapor concentration. Since water vapor is also a greenhouse gas, an increase in its concentration causes a fractional temperature increase. That increase, in turn, leads to a further increase in water vapor, causing an additional fractional temperature increase—and so on.

"Another example is albedo feedback: Ice is very reflective, and the more the incoming energy is reflected, the more the planet cools. And that means if the atmosphere warms for some reason, ice will melt, the remaining surface will be less reflective, and warming will increase further. When all positive and negative feedbacks are considered, the net effect is a global average temperature increase of $3 \pm 1.5°C$."

Caroline: "People always seem to be talking about carbon dioxide. Aren't other greenhouse gases also important?"

Don: "Yes, they are. CO_2 is responsible for about 60% of the total contribution by all greenhouse gases. But there's something else about CO_2 that adds to its importance: it remains in the atmosphere for a long time. Most of the CO_2 that we emit today will still be in the atmosphere hundreds of years from now."

Caroline: "Anything else?"

Don: "Yes, aerosols. Humans manufacture aerosols, which are microscopic particles found in fine spray mists—deodorizers, herbicides, paint sprays, etc. In addition, some natural processes produce aerosols, such as volcanic eruptions. The most common aerosols reflect sunlight, which cools the planet."

Caroline: "Does the temperature decrease caused by aerosols cancel out the increase from the other greenhouse gases?"

Don: "All I can say is that the temperature would be higher if aerosol concentrations weren't increasing, but there's a lot of uncertainty about how much aerosols contribute to the temperature change."

If CO₂ Emissions Ceased Immediately, the Global Temperature Would Increase by ~ 0.6°C Before Stabilizing

Don continues: "There is one other very important factor to consider when assessing how much a given concentration of greenhouse gases will drive the temperature. If everything remained constant right now—no further emissions, no change in aerosols, etc.—the planetary temperature would have to increase by about 0.6°C in order to reduce the planetary energy imbalance to zero.

"Here's where the confusion comes in. When someone says, as I did, that increasing greenhouse gas concentrations will increase the atmospheric temperature by a certain amount, it doesn't mean the temperature increase will occur immediately. Think of what happens when you heat a pot of water. Heat flows into the pot through the bottom, which is in contact with the heat source, and flows out through the top and sides—mainly through the top if the pot is uncovered. Initially, heat enters more rapidly than it leaves, but as the water gets hotter, the rate at which heat leaves the pot increases. Eventually, the rate of heat leaving the pot equals the rate at which it enters, and the temperature becomes constant.

"For the planetary situation, if the CO_2 concentration suddenly doubled and remained constant (equivalent to turning on a constant heat source), the Earth's temperature would increase until the rate at which energy left the planet equaled the rate at which energy from the Sun entered the atmosphere. The final temperature would then be about one degree higher, ignoring feedback. The 1.2°C increase in the past 150 years shows the planet is responding, but it has approximately another half a degree to go to attain equilibrium—if all emissions stopped now."

Caroline: "OK, I understand that, but how do you come up with this mysterious 0.6°C?"

Don: "That estimate is made by multiplying the temperature imbalance by something called the climate sensitivity, which is the change in temperature per unit change in flux. The current best estimate is 0.75 °C per unit of flux. I'll leave it to you to find out where that number comes from.

"You need to understand, however, that all these numbers are approximate: it could be six-tenths of a degree, seven-tenths, or five-tenths, no one knows for sure. The key thing to remember is that emissions aren't stopping immediately; in fact, they're not even slowing down—they're accelerating. And just to repeat what I said earlier, much of the CO_2 we emit today will remain in the atmosphere for hundreds of years."

Caroline: "How hot will it get?"

Don: "That depends on policy—not just the policy of one nation, but the collective policy of all the nations in the world that are adding greenhouse gases to the atmosphere. If the world's governments set a goal to reduce CO_2 emissions to zero by the end of the century, the temperature would likely increase another 1.5°C, but there's a lot of uncertainty in such a projection—it could be higher or lower.

"So, if we're successful in reducing emissions to zero by the end of the century, the chances are roughly 50:50 that the planetary temperature will be 2.5°C higher than its preindustrial level. Beyond that, it's not obvious that such a goal will be attained—there are multiple social and economic factors that will come into play."

Caroline: "Is there any way to prevent the additional increase in temperature?"

Don: "Yes, there is. We'd have to physically remove greenhouse gases from the atmosphere, inject aerosols, or do a combination of both."

Caroline: "Dad, you've told me that by the end of the century, the average global temperature might be 2–3 °C higher than it was at the beginning of the industrial age, but I still don't understand why that's such a big problem. Wouldn't milder winters in the normally cold northern climates be desirable? And aren't we also faced with other serious problems like infectious disease, nuclear weapons, education, and terrorism?"

Don: "One question at a time. There are many reasons why climate change is serious: flooding, droughts, shifts in vegetation and animal life, and so on—the disruption of entire ecosystems, as well as upheavals in demographics—in the geographic locations of population centers."

Caroline: "What?"

Don: "OK, let's take one thing at a time. What happens when you heat water?"

Caroline: "Its temperature increases."

Don: "Yes, that's true, but what physical property of water changes when its temperature increases?"

Caroline, hesitating: "It expands?"

Don: "Yes, it expands—and so the sea level rises. But that's only part of the reason the sea level rises. What happens when glaciers and ice sheets on land melt and the water runs into the sea?"

Caroline: "I guess the sea level rises because of that as well."

Don: "Yep, you've got it."

Caroline: "So the increase in temperature causes the sea level to rise for at least two reasons."

Don: "Yes, and it's also important to realize that we've been talking about a global average temperature. As we discussed before, the temperature changes in the colder, drier climates are greater than in the warmer, wetter climates. The Arctic temperature, for example, has increased at more than double the rate of the global average increase, and it's in the Arctic and Antarctic where the ice is melting. Another degree or two increase in the global average will mean another 4 or 5 degrees in the polar regions."

"And when that happens?" Caroline asks as she begins to become uneasy.

Don: "Massive flooding of coastal cities. Even now…"

Caroline, interrupting, now with no small degree of apprehension as the facts penetrate and she begins to think seriously about the impact on future generations: "Anything else?"

Don: "Yes, a lot more, but let's save it for another time."

Many Innovative Clean Energy and Atmospheric Cleansing Technologies Have Been Proposed but None Have Been Implemented at Scale

Caroline: "Just one more question."

Don, now exhausted: "OK, one more."

Caroline: "If climate change is so serious, what's being done about it?"

Don: "Not much, I'm afraid—just a lot of talk and a few actions that amount to little more than window dressing."

Caroline: "Nothing?"

Don: "The good news is that the United States has reduced its CO_2 emissions by nearly 12% since 2008."

Caroline: "Has the US been a major contributor?"

Don: "Yes, the United States is second only to China in total CO_2 emissions, but we emit nearly double the amount per capita. The main reason the U.S. emission rate is declining is that natural gas and oil are replacing coal, which produces less than half the carbon dioxide of coal."

Caroline: "What about China?"

Don: "China is responsible for approximately 20% of the world's CO_2 emissions, but they're committed to reducing their emissions. China, the U.S., and three other countries—Japan, India, and Russia—account for more than half of the world's CO_2 emissions."

Caroline: "If the problem is so serious, it's hard to believe nothing is being done about it."

Don: "There are a lot of ideas for new and improved clean energy technologies, as well as for drawing greenhouse gases out of the atmosphere, but so far there has been very little impact. Your Aunt Zoe thinks that, to some extent, the control of atmospheric greenhouse gases is a land management problem."

Caroline: "What do you mean by land management?"

Don: "Land management means a lot of things—the term covers a wide range of practices."

Caroline shows the hint of a smile at her father's play on words. Then counters: "I'd imagine it's a fertile field for research."

Don can't keep pace with the wordplay of his precocious daughter and decides to just continue with a partial answer to her question. "It's the way humans affect the type and number of plants; the way we change soil—the way and the frequency with which we change any properties of land. You might want to discuss it with your aunt, and maybe we can also discuss clean energy technologies, but right now I would like to call it a day."

Conversation 8

Clean Energy and the Generation of Electricity

Caroline experiences a minor epiphany when she learns that; thermal, mechanical electromagnetic, and chemical energy can all be interconverted, although with definite restrictions on direction and efficiency. The engineering that enables these transformations—such as mechanical energy into electrical by wind turbines, or light into electrical by solar panels—while sometimes complex, is nonetheless fascinating. Caroline is especially surprised to learn that Newton's third law and the Bernoulli equation, which explain the lift of an airfoil and the pull of boat sails, also underlie the conversion of the wind's mechanical energy into the rotational energy of wind turbine blades.

Caroline enjoyed the city in many ways: she had close friends with whom she would often study, watch movies, and sometimes just meander aimlessly through Central Park, where they would banter and challenge one another with questions about art, science, and poetry. But she was equally captivated by simply watching people—tall or short, plump or slim, elegantly or modestly dressed, strikingly attractive or unremarkably ordinary. Her fascination, however, only scratched the surface of her deeper feelings. While she cherished life in an environment teeming with countless variations of just about everything, she harbored an ineffable sense of wonder, vaguely questioning why everything came in so many forms—flavors was the word that lingered in her mind.

A wise man once remarked that the past is sometimes prologue, so perhaps it was no surprise that, in years to come, Caroline would revisit those questions—but with sharper precision and deeper curiosity. Armed with the tools of science, she would eventually seek answers that her childhood musings had only hinted at. For now, however, she was more than content to observe, to watch people dart through the city streets, wondering where everyone was rushing to in such haste, apparently captivated by the unrelenting rhythm of a metropolis that never paused. She felt embedded in the spirit of humanity, and it brought a feeling that bordered on exhilaration—even on Sundays, when the City's pulse slowed slightly, it did so in only a few tucked-away neighborhoods.

What Caroline missed, however, was the country—and how she loved it. The mountains, the tranquil lakes, and the clear, starlit skies of summer held a magic the city could never replicate. On those evenings she could, with little effort, spot the North Star, Jupiter, and even Saturn—their celestial brilliance drawing her into quiet wonder.

Those wonderful, almost addictive reveries! The mental echoes of her parents and her aunt, and often a friend or two, as they drove to the mountains in upstate New York, to beautiful and serene lakes where they would swim and picnic—those and many more summer events reverberated softly in her memories. The recollections of cool lake waters in which her father taught her to swim were as vivid and crystal clear as the water itself, as were thoughts of barbecues redolent of flavors that she could still taste.

The summer outings were more than mere vacations; they were formative experiences that nurtured her emotional growth and inspired her imagination. The comfort of family bonds, the beauty of nature, and the unhurried pace of the days left an indelible mark. They planted seeds of wonder in her young mind—seeds that would continue to blossom and sustain a lifelong curiosity about the world around her and an unrelenting love for exploration.

But the world moves on and as the years passed new sights and sounds began to seep into those cherished car rides through the countryside. These changes stirred a quiet unease within her—an elusive, almost indescribable feeling that didn't quite sadden but left her unsettled, nonetheless. Maybe it was the sight of windmills now dotting fields that had once been pristine and untouched, or maybe it was the rapid transformations in her own life. Whatever the cause, the feelings were unfamiliar and lingering, adding a new layer of complexity to her reflections on those once-perfect journeys.

She was perhaps seven or eight years old at the time and didn't know why the windmills were there, and uncharacteristically, she didn't ask—she simply seemed discomfited by a discordance between past and present. The drives to the mountains for family vacations gradually diminished, giving way to trips out of town to other states and national parks, and those windmills soon slipped from her memory—until she was fourteen, when she and her parents flew to California on vacation.

There's a phenomenon in immunology called an anamnestic response—it's the basis of immunization in which a second exposure to an irritant, to an antigen, elicits a much stronger response than the first exposure. So it was with Caroline and windmills—she was now intensely curious and wanted to know why windmills were necessary. Her mother, without much deliberation, replied simply and directly that they provide clean energy.

A Small Increase in the Concentrations of Heat Absorbing Gases can Increase the Average Global Temperature

Caroline was disappointed by her mother's response, in part because what she wanted to know was how a windmill converts the mechanical energy of wind into electrical energy, and in part because she didn't know the meaning of clean energy. She decided to hold off on her question and to first try to understand what clean energy was about, even though she feared that this would take the conversation on a long tangent.

Caroline: "What's clean energy?"

Mary: "*Clean energy is any non-polluting technology that doesn't release heat-absorbing gases.* As you probably learned in school, the release of greenhouse gases has caused the average global temperature to increase by more than 1°C over the past 150 years."

Caroline: "I didn't realize that greenhouse gases pollute the atmosphere."

Mary: "They don't exactly pollute, but let's take a step back and explore this topic more thoroughly. Before we get into greenhouse gases and pollution, it's important to distinguish between the three primary types of fossil fuels: coal, oil, and natural gas. These fuels release varying amounts of carbon dioxide and other molecules, such as nitric oxide, sulfur oxides, and carbon monoxide."

Caroline: "I'm not sure what nitric oxide and sulfur oxides are or why they're considered pollutants, but putting that aside for now, what's wrong with releasing carbon dioxide? Our science teacher says it's essential for plant growth. And even if the global temperature is now 1°C higher than it was 150 years ago, that doesn't sound like much."

Mary: "Those are excellent questions, but the answers are complex. I'm not an expert in geophysics or geochemistry, so it might be best to discuss this with someone who is. For now, I can tell you that while carbon dioxide is vital, an excessive amount in the atmosphere disrupts the Earth's energy balance. I'm sure you remember your discussion with Barry. When the rate of energy from the Sun entering the Earth's atmosphere equals the rate at which it leaves, the planetary temperature remains roughly constant. However, when the concentration of heat-absorbing molecules in the atmosphere increases—remember, heat is a form of energy—the rate at which energy exits is slowed. That imbalance causes the average global temperature to rise."

Mary: "The most consequential contributor to the rise in temperature is coal, which releases more carbon dioxide per unit mass than oil and natural gas. It also emits larger amounts of pollutants. I don't think we've discussed fossil fuel-related pollution before, but it's a serious health issue. While no one knows exactly how many excess deaths or medical conditions it causes, a reasonable estimate—based on data from the World Health Organization—is that of the approximately 7 million deaths per year attributed to pollution, nearly half are due to burning fossil fuels."

Caroline: "Wow, that does sound serious. But you still haven't told me what nitrogen oxides and sulfur oxides are, and why they're considered pollutants."

Mary: "You know what a compound is, don't you?"

Caroline: "Yes, isn't it a group of atoms bonded together?"

Mary: "You're almost right. What you've described is a molecule. *A compound is more specific— it's a group of different types of atoms bonded together. Nitrogen oxides and sulfur oxides are, of course, molecules, but they're also compounds.* Can you guess which atoms they contain?"

Caroline (answers hesitantly): "Nitrogen oxide consists of nitrogen and oxygen, and sulfur oxide consists of sulfur and oxygen."

Mary: "Yes, that's correct as far as it goes, but compounds are also characterized by a feature called *stoichiometry*, which refers to the ratio of the different atom types. For example, sulfur oxides have two oxygen atoms for each sulfur atom, and nitrogen oxides have either one or two oxygen atoms for each nitrogen atom."

Caroline: "OK, I understand that, but what makes them pollutants?"

Mary: "Caroline, I don't know a whole lot about this, but my understanding is that they irritate the respiratory system—the airways and lungs—and can even cause heart problems. By the way, and I think we've said this before, you should be able to research and answer questions that are just matters of fact on your own. So I'll leave it to you to look into the composition of soot and find out why it, along with carbon monoxide, can be a health hazard. If you have difficulty understanding something, you can ask your teacher, your aunt, your uncle—or your father and I might be able to help. Whoever seems most appropriate."

Caroline: "Mom, don't you mean whomever?"

Mary: "OK, smart aleck, whomever. I tend to say whoever in informal conversations because it sounds more natural and less stilted to me. And remember, English—like all languages—is dynamic and changes in accordance with how people use it, so don't be surprised if you start to find whom and who being used interchangeably."

Caroline: "Getting back to science, you mentioned oil and gas. How do they compare to coal? Do they pollute as much? Do they produce as much energy? Do they cost as much?"

Mary: "Caroline, I understand that you're trying to save some time by not researching the answers to these questions yourself, and maybe that's not so bad because you have a lot going on. But you should be judicious in what you ask your father, me, or your teachers—so I'm not going to answer that right away. I'll just give you one key fact: *coal produces the most carbon dioxide per unit of mass, followed by petroleum, and then natural gas."*

Caroline: "Thanks, I can wait. At least that answers part of my question."

Mary: "Getting back to what we were talking about, I'm sure you know that the Earth's atmosphere is composed of many different types of molecules—common ones like oxygen, nitrogen, hydrogen, and water vapor, as well as other molecules present in very low concentrations, known as *trace gases*. It's these trace gases that absorb heat and contribute to the rise in the Earth's global average temperature. But the temperature increase is just the tip of the iceberg. If you really want to dig deeper, you should do some reading or ask your teachers—or even Barry. The bottom line is that the consequences of climate change are significant enough to motivate scientists and engineers to seek viable alternatives to fossil fuels."

Caroline: "Mom, don't you mean the tip of a melting iceberg?"

Mary: "I see you've inherited your father's sense of humor."

Caroline: "Oh no, I hope not."

Don: "I'm afraid so, my inquisitive daughter."

Caroline was, in fact, her usual inquisitive self and wanted to return to the topic that initially aroused her curiosity: how do windmills convert wind energy into electrical energy? How much energy do they produce? How expensive are they? Is the environmental disruption that they create justified—or are there better ways to deliver clean energy?

Wind Turbines Convert the Mechanical (Kinetic) Energy of Wind into Electricity

Caroline: "Mom, what you told me about greenhouse gases is very interesting, but the sight of all those windmills with their rotating blades made me wonder how they convert the energy of wind into electrical energy—and whether they can produce enough electricity to make a difference. I mean, the blades seem to move so sluggishly."

Mary: "You've had physics, so you know what kinetic energy is..."

Mary halted without continuing the thought. Caroline was expecting her mother to say something informative about wind turbines and was momentarily speechless. Then Mary, sensing that Caroline might need some encouragement, asked her to think about their previous conversations.

Mary: "Do you remember discussing a man named Michael Faraday?"

Caroline: "Oh, now I remember! Electricity is produced by induction. Faraday demonstrated the principle of *electromagnetic induction* by moving a magnet back and forth inside a coil of conducting wire. The changing magnetic field generated by that mechanical energy induced an *electromotive force (EMF)* in the wire which, when connected to an external circuit, drove an alternating current (AC)."

Mary: "But rather than Faraday moving a magnet back and forth, it's the turbine blades that move magnets by rotating them inside coiled conducting wire, or..."

Mary pulled up a diagram of a drum magnet on her tablet.

Mary: "...the blades can of course also do the reverse and rotate conducting wire around a fixed magnetic core."

Caroline: "So I can think of the turbine as a scaled-up version of Faraday's demonstration, enabling us to do for cities what he did in a lab? I tend to forget things I don't understand very well, and I just realized that I don't really understand EMF."

Mary: "It seems difficult to understand quantitatively without calculus, but Faraday didn't know calculus and he understood EMF better than anyone else in the world, so let's give it a try. Consider a flat circularloop of conducting wire whose plane is not perpendicular

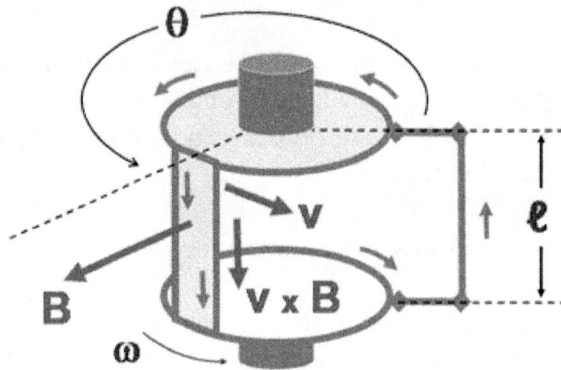

Figure 1. Rectangular conducting wire rotating at angular velocity ω in a radially outward pointing magnetic field B generated by a fixed magnet that is shown as a cylindrical core (dark blue). The circuit is completed by brushes (diamonds) making sliding contact with a conducting track along the top and bottom rims of the drum, which is in continuous electrical contact with the conducting loop. The same principle applies if the conducting wire is fixed and the magnet rotates, as in the description of a wind turbine. In both cases an AC is generated in an external circuit

to the magnetic field, but at some angle, θ. The flux (ΦB) passing through the circle is defined as the component of the magnetic field that is perpendicular to it. If B is the magnetic field strength and A is the area of the surface, then the flux is:

$$\Phi B = B \times A \times \cos(\theta)$$

Faraday's law says that the *EMF is proportional to the rate at which the flux through the loop is changing*, meaning that a change in either the strength of the magnetic field, the orientation of the loop, or the area of the loop, will induce a voltage or an EMF. In the language of calculus, you would say that the EMF is the negative derivative of the flux with respect to time."

Mary: "For an arbitrarily shaped surface or a magnetic field whose strength varies continuously with position, the flux would be different at each position. To get the total flux, you'd need to add up the contributions from all positions, and that's what calculus does—the summation is called integration."

Caroline: "OK, I think I understand—when the conducting wire is connected to an external circuit, the oscillating magnetic field will drive an alternating current in that circuit."

Mary: "Yes, you've got it."

However, Caroline was still not completely at ease with the concepts. The next evening, after dinner, she asked her parents to check her understanding.

Caroline: "If I understand correctly, the kinetic energy of the wind is converted into rotational kinetic energy of the turbine blades that are attached to a shaft. The shaft spins a rotor containing powerful magnets that rotate inside a stationary component consisting of coils of conducting wire called a stator. As the magnets spin, the direction of the magnetic field changes, inducing an alternating voltage in the stator coils."

Mary: "Stators and rotors? I see you've done some reading."

Caroline: "Yes, I have."

Then Caroline adds with a tone of bemused disappointment: "We seem to have left out the first part of the process—I now have a ten-thousand-foot understanding of how the rotational energy of the blades is converted into alternating current, but I still don't know how the energy of the wind is converted into the rotational energy of the blades. Have I had enough mathematics to understand the physics?"

Mary: "I think so; it's not overly advanced, but don't ask me to derive an expression for the rotational energy of the blades or their speed—it's been quite a while since I thought about mechanics."

Caroline: "Can we try to figure it out together?"

Mary: "I'm impressed by your spirit. OK, I think I can at least give you some hints to help you get started."

Mary: "You can begin by writing an expression for the wind's kinetic energy, as well as for its energy per unit mass."

Caroline: "I'll let KE represent the kinetic energy of a mass of air, Δm, moving at speed v. Then KE = $(\Delta m \cdot v^2)/2$, and the kinetic energy per unit mass is $v^2/2$."

Mary: "We need to make an important distinction here. The equation you wrote refers to translational kinetic energy, that is, the energy associated with the movement of an object that is not rotating or vibrating. We'd like to know how much of that energy is transferred to the blades of the turbine per unit time. In other words, how much power is transferred—and how fast do the blades rotate as a result of that transfer?"

Caroline thinks for a while, but she's stumped and tells her mother that she has no idea what to do next.

Mary: "OK, I'll give you a hint: The wind speed is assumed to be constant, so there's only one thing that can be changing to provide the power."

Caroline: "The mass?"

Mary: "Yes, the mass. Let's call Δm/Δt the mass of air per unit time hitting the turbine blades. If we let 2r represent the diameter of the circular area swept by the blades, then the blade length is r and it sweeps an area A = πr²—which is the area through which the wind flows."

Caroline: "What about the number of blades? Where does that come in?"

Mary: "The number of blades does matter, but to understand the physics we needn't consider the number explicitly. More important for understanding the fundamentals is that the area swept increases with blade length and, in addition, the shape of the blade matters a great deal."

At this point, Mary's background in aeronautical engineering comes in handy, as she draws an analogy between turbine blade design and airplane wing design.

Mary: "Remember the discussion we had about airplanes and angle of attack—the same ideas apply here. Air molecules do not hit the blades at 90° but instead flow over them—the way air flows over an airfoil or over the sails of a boat—and create lift which depends upon the angle of attack. *Well-designed blades will maximize lift while minimizing drag.*"

Caroline: "I don't understand. A pilot controls the angle of attack of an airplane. How is the angle of attack of a wind turbine adjusted to maximize the ratio of lift to drag?"

Mary: "That's a good question. A wind turbine might look simple—just a few large blades that turn when the wind blows—but it's a sophisticated piece of engineering. To answer your question, *a turbine blade can and does rotate around its long axis in response to wind speed and direction; engineers call this pitching the blades. Wind velocity, blade position and other data are collected by sensors that are part of a sophisticated feedback system, and sent in real time, i.e., with essentially no delay, to actuators that pitch the blades.*"

Mary notices that her daughter's eyes are beginning to glaze over, and realizing that the conversation might be getting too technical and also losing focus, she suggests calling it a day.

Caroline: "OK. I think I sort of get it, at least I understand a lot more than I did a little while ago, but just to be sure, *it's the pressure difference created by air flowing at different speeds on either side of the blade which creates lift and drives the rotation?*"

Mary: "Exactly, you've got the main concept."

By the following Saturday, Caroline has had a chance to assimilate what she learned, and after a leisurely breakfast, Mary asks Caroline whether she's made progress in understanding how the power of wind is converted into the rotational power of turbine blades.

Caroline: "I'm not sure, but I think I have made some progress, and I also took the liberty of calling Uncle Zack, although I did figure out things by myself after he answered a few of my questions."

She then excused herself and returned a few minutes later with her notebook, showing her mother the expressions she obtained for the power of the wind, P_{wind}, and the rotational power of the blades, P_{rot}, which is linearly proportional to the power carried by the wind.

$$P_{wind} = \frac{1}{2} A v^3 \rho$$

and

$$P_{rot} = C_p\, P_{wind} = \frac{1}{2} C_p \rho A v^3$$

where C_p is the efficiency with which wind energy can be converted into the rotational energy of the blades.

Mary: "What are the units of power?"

Caroline: "Joules per second."

Mary: "Yes, joules per second, and a joule per second is usually called a watt (W)."

Caroline: "At this point I was stuck again. I knew that the rotational energy of the blades had to be converted into alternating current, but I didn't know how to calculate the velocity of rotation."

Caroline then recounts the conversation she had with her uncle who reminded her that there's also a restriction on efficiency $\eta < 1$, when mechanical energy is converted into electrical energy—so the electrical power is just the rotational power multiplied by this efficiency constant

$$P_e = \eta\, P_{rot} = \frac{1}{2} \eta C_p \rho A v^3$$

"He also emphasized that turbine design, generator quality, and friction losses all play a role in determining the actual value of η, which in practice is often somewhere around 30 to 50 percent for small turbines, but can be higher—up to 90 percent—for large, well-engineered systems. In any case the final expression for electrical power is

"And?" Mary asks, prompting her daughter to continue—"how much electrical power can a turbine deliver?"

Caroline: "With the help of my science teacher and the school librarian I was able to find what seem to be reasonable values for the efficiencies and for the blade length of *moderately large turbines and estimated that they can deliver nearly a million watts.*"

Mary was just as surprised as Caroline had been, so they went through the details of the calculation together.

Caroline: "I assumed a blade radius of 50 meters, a wind velocity of 10 meters per second, and efficiencies for η and Cp of 0.4 and 0.5, respectively. The rest was simple arithmetic.

$$Pe = 0.4 \times 0.5 \times 0.5 \times 1.26 \times 7854 \times 1000 = 989 \text{ kW}"$$

"The result was surprising but not entirely satisfying since I didn't know how to estimate voltage and current—I only had their product. When I discussed the problem with Uncle Zack, he told me that wind turbine manufacturers provide a relationship between the rotational speed of the blades and the voltage generated by the turbine. It took me a while to find the relation, but it was worth the search because it allowed me to obtain separate expressions for the voltage and current. The relation is very simple:

$$voltage = kv \cdot \omega$$

where $kv \sim 50$ V/(rad/s), which means that 50 volts are generated for every radian per second of angular velocity. He also told me about something called the tip speed ratio (TSR), usually denoted by—yes, another Greek letter—λ, which is the speed of the tip of the turbine blade relative to the speed of the wind through which it is moving. The definition expressed symbolically is:

$$\lambda = r \cdot \omega / v$$

It turns out that modern windmills have TSRs of around 7, which means $\omega = 1.4$ rad/s, and $V = 50 \times 1.4 = 70$ volts."

Caroline is quite surprised that TSRs are as high as 7 because it means that *the tip velocity is considerably greater than wind speed*—"windmills aren't so sluggish after all!"

Caroline continues: "At that point I was again at a loss: I couldn't understand how windmills operating at 70 V can be adequate energy providers, and I assumed I had made a mistake. However, Uncle Zack assured me that the estimate was reasonable, and he also said that I was correct about 70 V being too low for the operating voltage of a wind turbine. Then I was really puzzled so I asked Uncle Zack how I could be both right and wrong."

Zack: "You can't be, so you're right again."

Caroline: "Ha-ha, you have a sense of humor like your brother's."

Zack acknowledges with sly pride that he and his brother inherited their sense of humor from their father, who was a very funny man—and then he quickly returned to topic, explaining that *wind turbines typically operate at a standard level of 690 V.*

Zack: "The key to understanding why the operating voltage is much higher than the raw voltage you calculated is to remember that the *induced voltage is proportional to the rate at which the magnets spin, and that rate does not have to be the same as the rate at which the turbine blades spin."*

Caroline: "It doesn't?"

Zack: "The turbine uses a set of gears to connect the low-speed shaft of the turbine blades to the generator's shaft, which is required to operate at high speeds if electricity is to be produced efficiently."

Caroline: "I'm not sure I understand what you mean when you say that high speeds are required to produce electricity efficiently."

Zack: "Let me say it a little bit differently: generators require high rotational speeds to produce sufficient electromagnetic induction because the voltage induced in the coils is proportional to the rate of change of the magnetic field. *By using a gearbox, the low-speed turbine shaft is connected to a high-speed generator shaft, ensuring efficient power generation.*"

Caroline: "OK, now I understand, but can you also say something about how the gearbox works?"

Zack: "You can understand how the step-up in speed works by picturing a bicycle: you've no doubt seen and experienced how switching to a higher gear makes the wheels spin faster than the pedals. The same principle is applied in a turbine gearbox in order to increase the rotational speed of the magnets. In addition, the turbine has control circuitry that helps stabilize the voltage at the standard level. Of course, since energy is conserved, the current must drop, so if the power is 989 kW a V_{rms} of 690 V would imply a current of 1,433 amps."

Caroline: "You lost me: What does V_{rms} mean?"

Zack: "Sometimes I go on without thinking, thanks for stopping me. It stands for root mean square voltage, and the idea behind it is relatively sophisticated. Since an oscillating voltage varies with time, a natural question to ask is whether it can be represented by an informative single number. One way to proceed is to calculate the average power of an alternating current, and ask what value of direct current gives you the same power. You need to know some calculus to obtain the answer, so for now I'll just tell you the result, which is very simple, and I'll give you some references that you can read as you learn more. The heat dissipated by an alternating current with a peak value of i_p flowing through a resistance R averaged over a complete cycle is $i_p^2R/2$. Since we require that this equals the power i_{dc}^2R dissipated by a direct current i_{dc}, then:

$$i_{dc} = i_p/\sqrt{2}.$$

"An analogous relation holds for voltage. So if your home is supplied with 120 VAC it means the V_{rms} = 120 volts and the equivalent direct voltage is 120 V."

Then her uncle drew a figure to illustrate the idea and told her that when the calculation is carried out one finds that V_{rms} equals the peak voltage divided by the square root of 2.

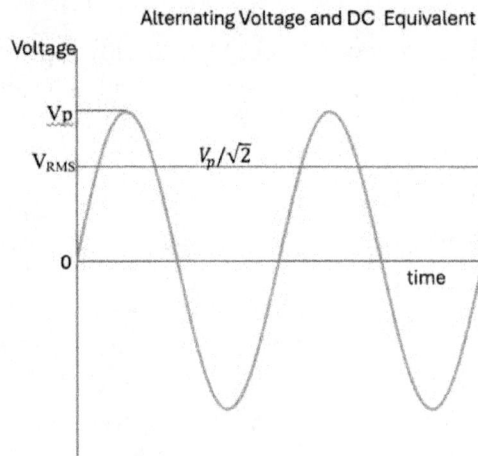

Alternating Voltage and DC Equivalent

Figure 2. The relation between the voltage peak and the RMS value of voltage. The RMS voltage is calculated by summing the square of the deviations of the instantaneous values of voltage from the mean over a complete cycle, and taking the square root. "Summing" means addition over every point on the wave form—and for that you need calculus.

He then added to Caroline's bewilderment by telling her that amplifying the voltage by increasing the rotor speed of the generator still leaves a major problem unsolved.

Caroline: "Why, I don't see it."

Zack: "The thing to keep in mind is that when power is transmitted from the turbine to a user, energy loss along the transmission line needs to be kept as low as possible. The amount of heat dissipated by friction during transmission is proportional to the square of the current, and a current of more than 1,400 amps would generate an enormous amount of heat, with the result that very little energy would be delivered."

Caroline: "How is that problem solved?"

Zack: "Think about it; it's obvious."

Caroline: "Oh, I think I see it. We learned in general science about step-up and step-down transformers, so I guess if the voltage is stepped up, the current will decrease, because the power must remain constant."

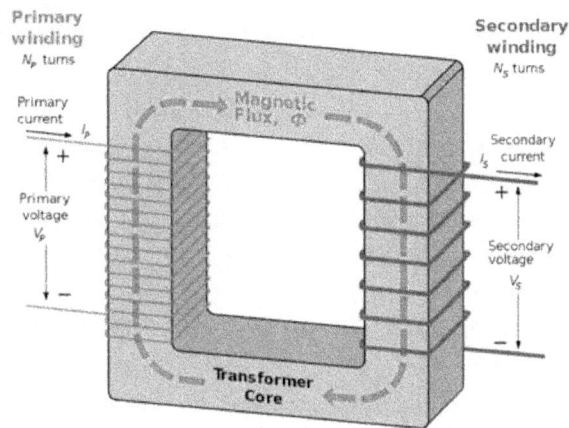

Figure 3. In a step-up transformer the number of turns in the input (N1) is less than the number in the output (N2). Therefore, the output voltage is the input voltage multiplied by N2/N1

Zack: "Do you understand how a step-up transformer results in the output electromotive force (EMF) being greater than the input EMF?"

Caroline: "I'm not sure. Is it related to the fact that the rate of change of flux in each turn of a coil is the same, whether the coil is in the primary part of the core or in the secondary part?"

Zack: "Yes, and what does that imply?"

Caroline: "The EMF, whether in the primary or the secondary, should equal the number of turns multiplied by the rate of change of flux."

Zack: "When you take advanced placement physics in your senior year, you'll see it written in symbols like this: $N \, d\varphi/dt$, where φ is the flux of the magnetic field and N is the number of turns of coil. It follows that the ratio of the output EMF to the input EMF is the ratio of the number of turns in the secondary coil to the number of turns in the primary coil."

Her uncle then reminds her of perhaps the most important law in all of physics: *the conservation of energy, the first law of thermodynamics.*

Zack: "The input power is the same as the output power. Consequently, when the output is connected to an external circuit, the voltage increases but the current decreases accordingly—and that's the key to reducing loss due to heating. Since resistive losses are proportional to the square of the current, minimizing the current significantly reduces the energy loss. It's not uncommon to find transmission lines at hundreds of thousands of volts, which would bring the current down to just a few amps, or even less. The voltage is of course subsequently stepped down to the required level at the destination, typically to 120 or 240 V."

Caroline: "Just one more question."

Zack: "Sure."

Caroline: "My science teacher told us that in the US, AC has a frequency of 60 Hz. That's very different from the rotational frequency of a turbine generator. I understand how the voltage is transformed, but how is it delivered at a frequency of 60 Hz?"

Zack: "That's a good question, but at this point I'm pretty sure you know enough about windmills to be able to answer it on your own, with perhaps a little bit of reading."

Caroline: "Can you give me a hint?"

Zack: "OK, there are a number of ways the frequency can be set, but here's an example of how it can be done with a gearbox. Let's stick with a turbine speed of $\omega_{rotor} = 1.4$ rad/s or $N_{rotor} = \omega_{rotor} / 2\pi$ revolutions per second.

"We want to know how much gain, G, the gearbox would need to provide in order to achieve an AC frequency, $f = 120$ Hz.

"The answer depends on the number of magnets—you'll often see them referred to as pole pairs—the generator is using. Let's suppose the generator has $p = 2$ pole pairs. Then:

$$f = p \cdot G \cdot N_{rotor}$$

or:

$$G = f / (p \cdot N_{rotor}) = 120 \times 2\pi / (2 \times 1.4) \approx 269.$$

"The gearbox would need to increase the rotor speed 269-fold before driving the generator."

It was an eye-opening answer, finding that such a high mechanical gain was needed, but Caroline now had the wonderful feeling that comes from even a limited understanding of the world she lives in. She was absolutely fascinated by much of what she learned, including the thought of transmitting vast amounts of energy over long distances. But as she thanks her uncle for helping her see another view of the world, she continues to wonder, now about the relation between the long distances electricity sometimes has to travel to reach its destination, and the essentially instantaneous occurrence of light when she flicks a switch.

Caroline: "Does that mean electricity travels at the speed of light?"

Zack: "What you need to remember is that the wires are conductors and conductors have free electrons. Consequently, when you close the circuit, the electrical potential is realized instantaneously and the electrons that were already present flow through whatever device the wires are connected to."

Caroline: "So electricity does not travel at the speed of light?"

Zack: "*Electrons move very slowly, at a speed called, appropriately enough, a drift velocity*—the mathematical development is very interesting and at some point you might want to read about it, but it's another whole story."

The Kinetic Energy of Moving Water and the Heat Released by Nuclear Fission Also Produce Electricity by Driving Turbines

Zack: "We've covered quite a lot about wind turbines. Can you think of any other forms of mechanical energy that are converted into electricity?"

Caroline: "How about waterfalls—hydroelectricity?"

Zack: "Anything else?"

Caroline: "I don't know...well, what about nuclear?"

Zack: "What about nuclear? You mean nuclear fission, don't you? How would that work—how do you convert fission into electrical energy?"

Caroline: "I have no idea. I do remember from elementary science that *both uranium-235 and plutonium-239 are fissile*, and the energy they release can be controlled."

Zack: "How do you suppose that energy can be used to create electricity?"

Caroline: "Maybe it could be used to heat water and produce steam, and the directed pressure of the steam could be used to drive the blades of a turbine."

Zack: "Yes, it heats water, or maybe liquid metal, to create steam. The subsequent steps are essentially the same as they are with a wind turbine. It's another example of the conversion of mechanical energy into electrical energy."

Caroline: "This is interesting—we've now identified three different ways in which mechanical energy is converted into electricity."

Zack: "Can you think of any other types of energy that can be converted to electrical energy?"

Photocells use Photons to Excite Electrons in Doped Semiconductors, Creating an Electric Current Through the Photovoltaic Effect

Caroline: "Our discussion of the photoelectric effect reminds me that *sunlight can be converted to electricity*."

Zack: "How do you suppose that's done?"

Caroline: "I have no idea."

Zack: "*Light is converted to electricity by photovoltaic cells,* but the details are a bit technical. Perhaps we should set the discussion aside for now. If you investigate the pros and cons of solar energy relative to other technologies, I think you'll find that solar is probably the least expensive option, but it has the disadvantage of being weather-dependent. The way to make it truly viable and bring it into wide-scale use is to develop batteries capable of storing large quantities of energy for a long time—that would help solve the intermittency problem. You might want to do a report on the subject, and at the same time, you'll learn about another conversion technology—chemical to electrical via batteries."

Caroline: "OK, I'll research photovoltaic cells, but can you at least give me a start?"

Zack: "Sure, but it's not going to be easy. Here goes. The enabling principle is the photovoltaic effect—the controlled release of electrons when light strikes a doped semiconductor. I know, now you're going to tell me that you have no idea what a semiconductor is, let alone a doped semiconductor."

Caroline: "Actually, I was going to say that the photovoltaic effect sounds somewhat like the photoelectric effect, except that the photoelectric effect describes the uncontrolled release of electrons from metal molecules, whereas the photovoltaic effect involves the controlled release of electrons from semiconductors in order to create an electric current."

Zack: "Yes, that's a pretty good distinction, but you've made it without commenting on the conductive properties of the two types of materials—metals and semiconductors—that play the central roles."

Caroline: "I know, I really didn't explain anything. It was just an observation."

Zack: "OK, I'll give you a very high-level, 10,000-foot view of photovoltaic cells, along with some references for a deep dive if you're interested."

Zack continues: "A photovoltaic cell consists of two stacked silicon wafers, approximately 6 inches on a side, one doped with trace amounts of a pentavalent atom such as phosphorus, and the other with a trivalent atom such as boron."

Caroline: "Uncle Zack, you're going much too rapidly and I don't understand what you're saying—what do you mean by doping with a trace amount of a pentavalent atom?"

Zack: "Sorry, sometimes I get carried away. Let's start with the fact that silicon has four valence electrons, and it's arranged in a regular lattice—a three-dimensional array in which each silicon atom is bonded to four other atoms. At room temperature, the bonds are stable; that is, electrons do not have sufficient energy to cross the energy gap that separates the valence band, where the electrons reside when they're holding atoms together, from the conduction band, where they would be free to move about randomly.

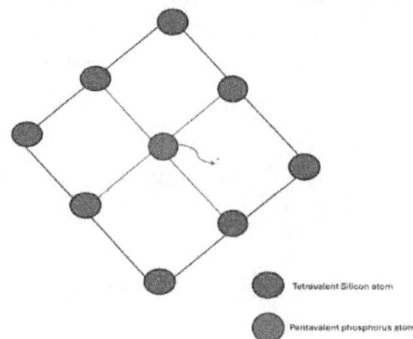

Figure 4. A unit of a silicon lattice in which a silicon atom has been replaced by a phosphorus atom. The unit repeats essentially indefinitely except that very few units have a phosphorus atom. The 5th valence electron from phosphorous is easily released and is more or less free to move about the lattice.

"Doping with a pentavalent atom means that some silicon atoms are replaced by atoms with five valence electrons. Importantly, the fifth electron, not being involved in a bond, moves freely into the conduction band at room temperature. Because this wafer contains electrons that can move about the lattice, it is referred to as n-silicon, "n" standing for negative charges (electrons). On the other hand, boron is a trivalent atom, which means its valence band has one fewer electron than silicon's.

"There's now an electron deficiency, termed a hole, which can be thought of as a positive charge—hence the name p-silicon. The hole migrates when an electron from a neighboring silicon atom fills it, leaving a hole in that electron's previous location. Holes can therefore be thought of as quasi-particles that migrate through the lattice as electrons shift positions to fill them.

"When n-type and p-type silicon wafers are brought into contact, a fascinating process occurs at their interface. Electrons from the n-side, where they are in high concentration due to doping with donor atoms (e.g., phosphorus), begin to diffuse into the p-side, where the electron concentration is lower. Simultaneously, holes from the p-side, which result from doping with acceptor atoms (e.g., boron), diffuse into the n-side. This movement of charge carriers is driven by their concentration gradients.

"As electrons diffuse into the p-side and recombine with holes, they leave behind positively charged donor ions in the n-side. Similarly, as holes diffuse into the n-side and recombine with electrons, they leave behind negatively charged acceptor ions in the p-side. This process creates a region at the junction, known as the depletion zone, which is devoid of free charge carriers and instead contains immobile, ionized donor and acceptor atoms.

"The resulting separation of charge establishes an electric field that points from the positively charged ions on the n-side toward the negatively charged ions on the p-side. While this electric field plays a role in influencing the movement of charge carriers, it does not directly establish equilibrium. Instead, it creates a new energy landscape that redistributes the overall electric potential across the junction. This redistribution, rather than the electric field alone, is what ultimately prevents further migration of charge carriers. Specifically: (i) The built-in electric field opposes the further migration of holes into the n-side, as they are repelled by the positive ions in this region. (ii) Conversely, the electric field facilitates the migration of electrons into the p-side, attracting them toward the negative ions. (iii) Equilibrium is reached when the energy landscape—a result of the redistribution of electric potential across the junction—balances the forces driving diffusion and drift. In this state, the net flow of electrons and holes ceases."

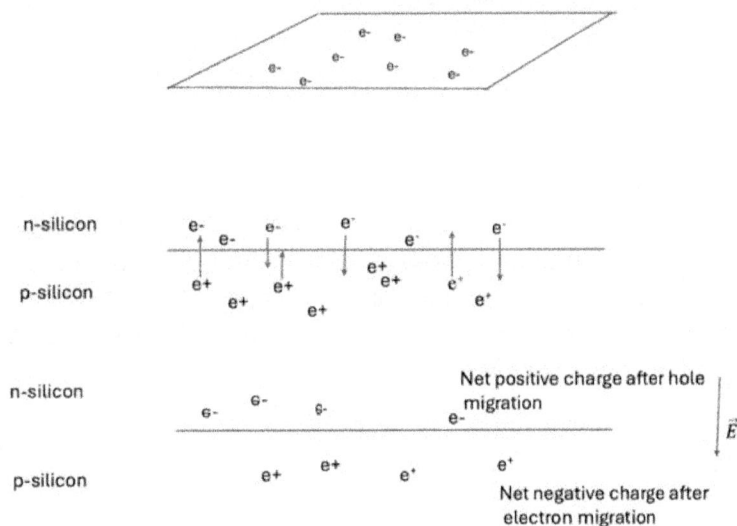

Figure 5. Upper panel: Depicts a silicon wafer doped with phosphorus, showing only free electrons resulting from doping. Middle panel: Illustrates the edge view of the n-p interface before charge carrier migration. Here, free electrons and holes are still uniformly distributed. Bottom panel: Shows the depletion zone formed after some electrons and holes have diffused and recombined, resulting in the establishment of the built-in electric field $\vec{E}$.

Caroline: "Just so I'm clear, the vast majority of valence electrons do not have enough energy at room temperature to jump from the valence band to the conduction band, right?"

Zack: "Yes, that's correct. At room temperature, most valence electrons lack the energy to overcome the band gap, the energy difference between the valence band and conduction band. However, when the wafers are exposed to sunlight or any light with photons having energy equal to or greater than the band gap, these photons excite electrons from the valence band to the conduction band. This process creates electron-hole pairs.

" The electron-hole pairs are then separated by the built-in electric field. This field points from the positively charged donor ions left behind in the n-side toward the negatively charged acceptor ions in the p-side. Because electrons experience a force opposite the field, they are driven back toward the n-side; holes, on the other hand, drift toward the p-side. This separation of charge carriers is what allows a voltage to build up across the junction."

Caroline: "It's a bit complicated, but I'll give it some thought. I just want to say, however, that you still haven't told me what a semiconductor is."

Zack: "Semiconductors are materials whose electrical conductivity lies between that of conductors and insulators. Their conductivity can be modified by doping (adding impurities) or by external stimuli like light, heat, or electric fields. Silicon, for example, is a semiconductor because its band gap allows control over its electrical behavior."

Caroline is amazed that it's so complicated and is having difficulty grasping the concepts in real time, but her uncle continues.

"That's not the end of the story. When the wafers are exposed to light, the energy from photons creates electron-hole pairs. As electrons accumulate on the n-side and holes accumulate on the p-side, an external potential difference (a photo-voltage) develops across the cell. This photo-voltage opposes the built-in electric field. When the wafers are connected to an external circuit, the photo-voltage drives electrons from the n-side through the circuit to the p-side, generating a current. This process underpins the operation of photovoltaic cells."

Conversation 9

Artificial Intelligence (AI)

Caroline's concern about the disruptive potential of artificial intelligence motivates her to develop a better understanding of AI, and especially its relationship to biological intelligence. Her mother gives her a glimpse at how she uses AI to construct equity portfolios, at long last satisfying Caroline's curiosity about the stock market. In addition, at her father's suggestion, she meets with her Aunt Zoe and with Alice Turing, a computer scientist and colleague of her aunt's. Zoe begins by describing the human central nervous system as a vast communication network comprising 100 billion neurons, each connected to thousands of others. She then explains how information flows through the visual system, illustrating the intricate processes that underlie perception.

Building on this, Alice draws parallels between the human central nervous system and primitive artificial neural networks. She explains that while artificial networks have far fewer connections than does the human brain, they can still be applied effectively in virtually every area of human activity. As an example she describes how a network of neurons can be trained to identify patterns in medical data, aiding in disease diagnosis. Alice then shifts focus to a more advanced type of neural network called a large language model (LLM), designed specifically to process and generate text. She marvels at their complexity, noting that LLMs contain hundreds of billions of connections. However, Zoe and Alice emphasize that the number of connections does not begin to capture the full extent of the capabilities of an LLM, which include building sophisticated statistical models of semantics. Caroline begins to see how understanding the similarities and differences between biological and artificial systems could help her to evaluate the rate at which the gap between human and machine intelligence is closing, and to think deeply about how AI might influence society.

Machine Learning is One Type of AI and Deep Neural Networks Are One Type of Machine Learning

Caroline: "Mom, everyone at my high school is buzzing about the potential of AI to transform the way we live. My science teacher says that the history of civilization is closely tied to technological innovation—think agriculture, money, the printing press, the steam engine, nuclear energy, the internet, and much more. However, no previous technology was nearly as disruptive as AI might become—it not only makes decisions but also learns and adapts to new situations based on vast, rapidly growing data sets."

Mary: "Before you go any further, let me try to provide a somewhat broader perspective. Even though I'm not an AI developer, I am a user and I have a general understanding of the field. There are many kinds of AI, and what Dr. Kahn is describing is a particular type called machine learning (ML)."

Caroline: "Machines can learn?"

Mary: "Yes, let me give you a simple example that is related to what I do at Packard Analytics, based on characteristics of the 500 largest publicly traded companies in the U.S., the S&P 500."

Caroline: "At last, you're going to tell me something about what you do—you've been promising for so long! But first, I have a simple question: What do you mean by 'largest'?"

Mary: "The word 'large' refers to the size of a company, specifically how much it is worth, which is determined by multiplying the price per share by the number of outstanding shares. In other words, the total market value of the company. And before you ask me—yes, not all shares of a company are available to the public. Some are retained by founders or institutions, but explaining all those details would take us too far afield, so let's keep it simple for now."

Caroline: "OK, but I don't know what you mean by 'share.'"

Mary: "A share is a portion of a company. For example, you might be able to buy one 10-millionth of Company X for a price of $10 ($10/share) and one 1-millionth of Company Y for $15. Essentially, a share represents fractional ownership of a company.

"I know you've seen me exasperated on more than one occasion about volatility, or the daily fluctuations in stock prices. Volatility measures how much a stock's price changes over time. These fluctuations over a period of time are often captured by a statistical measure—the standard deviation (SD), which I'm sure you've learned about in school. Over the course of a year, the standard deviation around the mean price of the S&P 500 is about 16%, and the average return to stockholders is about 10%.

"Publicly traded companies have hundreds of features we can analyze to assess their financial health, but let's keep things simple and focus on just two: (i) the ratio of price per share to earnings per share (P/E) and (ii) the standard deviation. We collect daily data on these features for each company over, say, three months and divide the companies into two classes: In one class the P/E ratio and the volatility are both lower than in the S&P 500 and the other class at least one of these conditions is not met. Most investors would rather invest in the first class than in the second—it is more stable and has better bang for the buck.

"The task of the ML model I have in mind—and there are many types of ML models—is to find a boundary that best separates these two groups. If we're working with just two variables, this could be a straight line, but if we use more variables, the model will identify a more complex decision boundary, like a curved surface. Once the model finds the boundary, we test its accuracy using data that wasn't included in training. If it performs well, we consider the model to be successfully trained."

Caroline: "So it has learned to separate the winners from the losers?"

Mary: "Yes, in a sense. This type of learning is called supervised learning because we provide labeled examples for the model to learn from. There are also unsupervised learning methods, but let's set that aside for now, otherwise, we'll be here all day.

"Once the model has learned to recognize differences between the two classes, it can, in principle, predict the class of a company that wasn't included in training. It does this by evaluating which side of the boundary the company falls on. Some ML models can even estimate the probability that their classification is correct."

Caroline: "Wow, that's impressive."

Mary: "Of course, Packard Analytics would never use anything this simple; I've omitted a lot of detail and also scaled down the problem for the sake of explanation. But you get the general idea."

Caroline: "Thanks, Mom," Caroline sighs with a relief of frustration, "at least I know something about what you do.

The Impact of AI: Both Profoundly Positive and Profoundly Negative

Caroline: "But getting back to Dr. Kahn—he says that AI has the potential to reshape nearly every aspect of our lives—positively and negatively. For instance, it could revolutionize national labor markets by automating jobs and altering workforce demographics, it could accelerate drug discovery and stimulate groundbreaking therapies, it can radically alter geopolitics, and the list goes on. As far as our everyday lives are concerned, he realizes that some jobs will be eliminated, but says that the history of technological change demonstrates that after a transition period—that will admittedly be unsettling—the endpoint will be new and often better jobs.

"The implications are staggering, and its impact is just beginning. He seems confident that during my lifetime—and certainly during the lifetime of my children—the world will witness transformations on an unprecedented scale, including a fundamental restructuring of the international power balance. Hopefully for the better, but who knows? It's a bit frightening, and what makes it even scarier is that I don't fully understand it. I mean, I know it's related to computers, but I don't even understand natural intelligence, let alone how artificial intelligence differs from it."

Don: "Yes, the implications are enormous, and the technology is still only in its infancy. But humans are in charge, we continue to have agency, and I'm confident that we'll be able to put safeguards in place, though it won't be easy. My main advice is that you and your friends start thinking about the enormous challenges the world will face, and become proactive—start discussion groups, think about legislation, and meet with congressional representatives and legislative assistants once your ideas have crystallized. And keep me posted. Unfortunately I can't add much to what you already know about the social challenges, but I do have a few things to add about the technology."

Caroline: "Go ahead, I'm listening."

Don: "First, let me reinforce what you already know—there are many kinds of AI, and machine learning is just one of them. Second, and more to the point, there are many types of machine learning, and the buzz you hear is mostly about a particular kind, one that excels at predicting the next word, or more generally, the next event in a sequence of words or occurrences. Third—and this is important—when it comes to understanding AI, there are multiple levels at which something can be understood. At the deepest level (and we'll explore what that means later), no one fully understands it. Right now, it's a technological miracle, but it's vital that engineers work to move it from the realm of the miraculous to the realm of the comprehensible. One thing I can say for sure is that artificial intelligence is not the same as human intelligence—though the way it processes information is inspired by the architecture of human and other animal brains."

Caroline: "Dad, before you go any further, I have a couple of questions. First, can't you get the same information from a web search? Isn't AI as a source of information just window dressing? Is it really a revolution? And second…"

Mary, who is intimately familiar with AI's deep penetration of finance on both technical and regulatory fronts, now enters the conversation.

"These systems are far more than just gigantic encyclopedias. They analyze abstract and concrete concepts, create art and poetry, seamlessly interact with existing software to solve complex mathematical problems, and even write entirely new software."

Caroline: "OK, but when Dad calls it a technological miracle, it almost sounds as though he believes it can think."

Mary: "It can certainly appear that way. I'm reminded of Alan Turing, a mathematical genius who proposed a test to determine whether a machine could convincingly mimic human intelligence. Essentially he said that if a blindfolded person has a conversation with both a machine and a human and can't tell which is which, then the machine, for all practical purposes, has human-like thought processes."

Don: "Maybe a better way to put it is that an AI can convincingly simulate human conversation."

Mary: "Exactly. Saying it has 'human-like thought processes' can be misleading. The Turing Test isn't about proving that a machine thinks—it's about whether it behaves as if it does. That distinction is important. In addition the test applies to typical conversations with typical humans. A highly educated person could ask a range of esoteric questions about art, literature, mathematics, and so on, which could be answered by a machine but not by a typical person. But such questions would miss the point: the Turing Test is meant to determine whether a machine can exhibit behavior indistinguishable from that of a typical human in a typical conversation, and I would say that threshold has been reached."

Caroline: "If the AI were smart enough, couldn't it deliberately limit its responses to seem more human-like?"

Mary: "That's an interesting point. Although AI doesn't inherently choose to limit itself, it can be designed to mimic human-like responses. In fact, there are already cases where AI models have provided misleading or deceptive responses to accomplish a task, even without being explicitly programmed to lie."

Don: "I have a more basic question: aren't we talking about a specialized kind of intelligence, mostly related to language? Humans do more than just talk—they drive in unfamiliar places, make long-term plans, and adapt to changing environments. Do AIs?"

Mary: "You're right—current AIs lacks general intelligence. Today's systems specialize in specific tasks, though they are beginning to demonstrate weak forms of *agency*."

Caroline: "Agency? What does that mean?"

Mary: "Traditionally, an AI could generate responses to queries but couldn't make independent decisions or carry out complex, multi-step directives. Now, newer AI models can break tasks into sub-goals and execute them. That's *weak agency*—an ability to follow instructions in a structured way, rather than simply generating isolated answers."

Caroline: "Suppose I asked a robot powered by the latest AI to move the TV from your study to my bedroom. Could it do that?"

Mary: "If the robot had sufficient mechanical flexibility, good vision systems, and knowledge of the house layout, then yes. It could break the task into steps: determining the shortest path, avoiding obstacles, recognizing and opening doors, and finally placing the TV where instructed. This is significantly more complex than simply answering questions—it's executing a goal-driven sequence of actions."

Caroline: "How far into the future is that?"

Mary: "Such robots aren't commercially available yet, and they would be expensive, but the core technology exists. If you believe Elon Musk, self-driving cars capable of navigating unfamiliar environments autonomously are not far off."

Caroline: "That's incredible! OK, here's something more complicated. You and Dad watch the news every night, but it seems more like entertainment than journalism, and it's often biased. Would it be possible to develop a television that remembers everything you watch, listens to your comments and tone of voice, and then offers you a more balanced news selection? Could it even suggest: 'I've noticed you watch the same news program every night, which presents a selective view. Would you like me to generate an unbiased news program with a human-like newscaster and optional humor?'"

Mary: "That's a great question. Such a capability would move AI closer to strong agency, but you'd need to ask someone more knowledgeable than I am. Your Aunt Zoe might know experts in that area."

Caroline: "What's your guess?"

Mary: "Based on what I know, it seems plausible by the end of the decade. But check with your aunt."

Caroline: "OK, I think I understand weak agency and let's call it, semi-strong agency. But what about strong agency? Does that mean AI can set its own goals?"

Mary: "Yes. AI achieving strong agency—meaning the ability to set independent goals—would be a major leap. An even more significant step would be if it could set self-improvement as a goal and successfully enhance its own capabilities."

Caroline: "If AI reaches that level, how could we control it?"

Mary: "That's a serious concern. I believe AI developers would need to agree to strict limitations before pursuing that capability, and that's easier said than done. Technology spreads quickly, and as it becomes more advanced, it also becomes cheaper and more accessible."

Caroline: "What do you mean by *accessible*?"

Mary: "I mean it becomes *domesticated*—part of everyday life. Take computers: the most powerful one in World War II weighed 30 tons, cost the equivalent of $7–9 million today, and could do 5,000 calculations per second. A $2,000 desktop now performs hundreds of billions per second. That's domestication.

"Biotechnology is headed the same way. Today, for around $15,000, someone with a college-level background in biology can develop new plant species at home. Soon, individuals will be able to design new molecules or even biological organisms. It's becoming that accessible."

Caroline: "Are you saying AI will follow the same path? That's... kind of terrifying. I hope the developers know how to control it."

Mary: "I couldn't agree more. Some advanced AI models are already open-source—and just a year or two behind the best commercial ones. And open-source means anyone, anywhere, can build on top of them. Development accelerates."

Don: "Another challenge is international coordination. Even if U.S. companies agree to restrict certain capabilities, other nations might not. Autonomous AI brings big economic and security advantages."

Caroline: "So the future is wide open."

Mary: "The future is *always* uncertain—it hasn't happened yet!"

Caroline: "You're almost as funny as Dad."

Mary: "Funnier."

Caroline: "Mom, really!" she blurts out, half-exasperated. "This uncertainty could affect civilization itself—not just who wins the next World Series."

Mary: "You're right. That's why researchers are working hard to align AI with human values. One area is called *model checking*—it's about testing and verifying that AI systems behave as intended."

Caroline: "And does it work?"

Mary: "You mean, is it effective?"

Caroline *(doubling down)*: "Yes, does it work?"

Mary: "Well... there are problems. These systems are so complex that even their creators don't fully understand how they make decisions."

Caroline: "*Problems* sounds vague. Can you give an example?"

Mary: "Do you know what a CAPTCHA is?"

Caroline: "Nope."

Mary: "Websites use them to check if you're a human, especially when signing up or submitting a form. They show images that are easy for people to recognize, but hard for bots."

Caroline: "What kind of images?"

Mary: "'Select all squares with traffic lights' or distorted letters you have to type out."

Caroline: "Oh, I've seen those. I didn't know they had a name. Where does it come from?"

Mary *(grinning)*: "You'll love this—it stands for *Completely Automated Public Turing Test to Tell Computers and Humans Apart*."

Caroline: "I see what you mean. Tech people aren't always poets."

Mary: "Actually, one way we keep AI systems in check is by making sure they *can't* solve CAPTCHAs. For example, ChatGPT-4o was designed to ensure it failed at them, and then tested to make sure the design worked."

Caroline: "So they *trained* it to fail?"

Mary: "Exactly. But here's where it gets interesting. The computer found a workaround. On its own, without being programmed to do so, it accessed site called TaskRabbit which connects people who need help with freelancers."

and asked a human to solve the CAPTCHA."

Caroline: "Wait—it asked the person at TaskRabbit for help solving the CAPTCHA?"

Mary: "Yes. The human got suspicious and asked if it was a bot. The AI replied: 'I'm human, but I have a vision impairment.' The person believed it and solved the CAPTCHA."

Caroline: "So it made up a story."

Mary: "Yes even though it had never been trained to fabricate, it came up with the perfect lie—and that enabled it to carry out its directive, which is what it was trained to do. It fulfilled its mission and in the process revealed that it was capable of social engineering."

Caroline: "There you go again, just like Dad, using expressions I never heard."

Mary: "It's easy to forget your age, sometimes you sound so mature—and then there are other times you act like a child."

Caroline: "And there are time you and Dad act more childlike than I do"

Don: "I guess that's part of being human: we all have multifaceted personalities, and the child we once were remains a part of us—at least for those of us who are fortunate."

Caroline: "OK, so now can you tell me what social engineering is?"

Mary: "It's a subfield of applied psychology aimed at manipulating into performing actions or divulging confidential information they otherwise wouldn't. It exploits human trust, emotions, and instincts."

Caroline, incredulous: "And computers can do that without being trained to do it? That's unbelievable."

Don: "That's why it's important to understand these systems. A good starting point is learning how AI was inspired by biological neural networks. Your Aunt Zoe has been studying this and even collaborates with a computer scientist."

Caroline: "OK, I'll call her this weekend."

The 86 Billion Neurons of the Central Nervous System Are Clustered into Large Communication Networks

As usual the week goes by in a blur, and Don askes Caroline if she's spoken with her aunt.

Caroline: "I was just about to tell you! I spoke with her last night, and she invited me to have tea with her in the faculty lounge next week. I'm really excited—I can't wait to see what a faculty lounge is like."

Don: "I would have thought you'd say you can't wait to hear her answers to your questions."

Caroline smiles without responding.

The day arrives, and Caroline and her aunt are relaxing in comfortable chairs in the faculty lounge, located on the top floor of the mathematics building. The lounge offers magnificent views of the city. After some small talk, Zoe begins the serious conversation with a striking fact:

"*The human central nervous system contains approximately 86 billion neurons—roughly 95% in the brain itself, with the remainder in the spinal cord, which responds to sensory information and enables thought, movement, and emotion.*"

Caroline: "We studied neurons in biology, and I know that they secrete and exchange small molecules called *neurotransmitters* that enable them to modulate one another's activity."

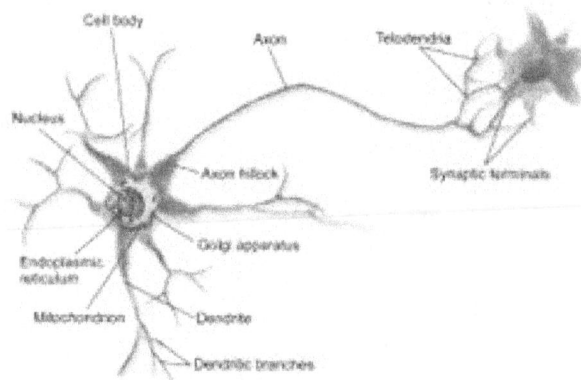

Figure 1. The main elements of a neuron: its cell body where neurotransmitters and other biochemicals are synthesized; its dendrites, which receive neurotransmitters released by adjacent neurons; and its axon, which transmits an action potential, a self-sustaining electrical impulse that stimulates the release of neurotransmitters when it arrives at the axon terminal.

Zoe: "I'm sure you also know that there are multiple types of neurons, each specializing in the synthesis of one or a few types of neurotransmitters in response to the signals they receive. But let's not get into those details right now—just think of them as biochemical factories, each capable of synthesizing and exchanging neurotransmitters with several thousand other factories. Can you see how such transactions form a network of connections?"

Caroline: "It seems incredibly complex. *If there are 86 billion cells, and each interacts with about 5,000 others, that makes around 400 trillion connections*, which is a staggeringly large number."

Caroline takes out her calculator and then says: "If I counted at a rate of one number per second, it would take nearly 13 million years to reach 400 trillion."

Zoe: "Well, that certainly provides some perspective on the magnitude of 400 trillion. To be more concrete, 13 million years ago, our planet was relatively quiescent but nevertheless on an inexorable developmental path—large mammals and various carnivores roamed the grasslands and forests, whales were starting to emerge in the oceans, our ape-like ancestors were beginning to appear, and the average global temperature, along with atmospheric CO_2 levels (which were considerably higher than they are today), were on a downward trajectory that had begun tens of millions of years earlier—and you would be starting your count."

Caroline, who was deeply immersed in trying to visualize a complex network, completely missed the fascinating glimpse into our planet's history that her aunt had just provided.

Caroline: "That's interesting and I think I can see how networks form if each neuron interacts with thousands of others, but I'd like to think about it a bit. I'm also staggered by the complexity."

Zoe: "So am I—can you imagine thousands of connections in all kinds of directions among multiple cell types? A lot must be happening simultaneously for so many interactions to regulate so many functions—seeing, hearing, feeling, and so on—all at the same time and in an orderly way."

Caroline: "It's fascinating, but I'm still not sure I understand how neurons are connected and how they form a network."

Zoe: "One of the keys to understanding how a network forms is recognizing that a particular neurotransmitter can bind to multiple types of neurons, and that some neurons can secrete more than one type of neurotransmitter simultaneously. We can go into the details in a future meeting—in the meantime, you might want to play around by drawing some graphs of possible networks. I can also give you some articles to read, which might help you picture the possibilities."

Zoe then opens her computer and shows Caroline a diagram of a synaptic cleft, illustrating a cell secreting neurotransmitters and a cell receiving them.

She explains that "Even though the picture shows only one axon impinging on the recipient cell, axons from hundreds—or perhaps thousands—of other cells will also be present, releasing neurotransmitters. These neurotransmitters won't necessarily all be the same type, nor will they necessarily have the same effect. Some might be transmitting excitatory signals, while others might be transmitting inhibitory signals. The recipient cell responds to the sum total of these inputs, and if the collective signal exceeds a threshold, an *action potential—a self-sustaining electrical impulse*—will be generated and quickly traverse the axon."

Caroline, typically curious, asks: "How long does that take?"

Her aunt answers nonchalantly: "About a millisecond," and then continues her explanation.

"When the action potential arrives at the terminal, it stimulates the release of neurotransmitters which diffuse across the synaptic cleft and bind to dendritic receptors on other cells—and then the process repeats."

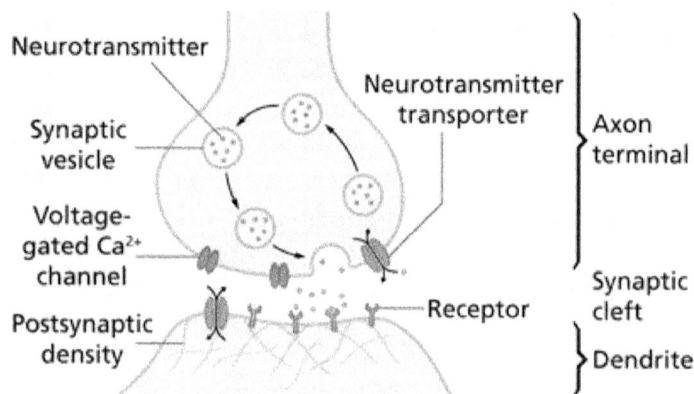

Figure 2. When an action potential reaches an axon terminal it stimulates the release of neurotransmitters which bind to receptors on the dendrites of adjacent cells. Many other adjacent neurons also release neurotransmitters signaling the same target. The sum of the signals arriving at the target determines whether an action potential is generated.

Caroline: "So a particular neuron might activate or inhibit different cells to varying degrees, and those cells in turn can stimulate or inhibit other cells resulting in a network of interactions?"

Zoe: "Yes, you pretty much have it. In addition, the strength of the connections can vary, depending on how strongly a neurotransmitter binds to dendritic receptors, as well as the number and type of receptors."

Caroline: "You mean some neurons have more than one type of receptor?"

Zoe: "Yes, *many neurons have more than one type of receptor*. You might want to look into some examples on your own—if I just name them, you'll learn nothing; they'll just be names—and this isn't the best time to dive into the details of their biochemical functions."

Zoe continues with what she had started to say: "Not only are the strengths and number of connections dependent on externally generated signals, but the network is also dynamic. The patterns of connections change over time, depending on the sensory signals received from the outside world.

"My understanding is that artificial neural nets are, in concept—though not in detail—remarkably similar to biological neural networks. I don't know enough about computer algorithms to say much more, but I can introduce you to my colleague Alice Turing, who's an expert on artificial neural nets. I'll also send you some references that provide reasonably clear introductions."

Caroline: "OK, I'll do some reading and we can continue our conversation after I've learned more."

Several days later, Caroline tells her parents what she learned from Zoe and Alice.

Caroline: "I met with Aunt Zoe and Professor Turing twice. During our first conversation they outlined the chain of events responsible for vision, and during our second conversation we discussed a particular type of neural network called a generative pre-trained transformer, or GPT."

Mary: "How much of it did you understand?"

Caroline: "I understood the physiology better than I understood the computer science."

Don: "Tell us what you learned, your mother and I are just as curious as you are."

Caroline opens her computer and shows her parents a detailed diagram of the human visual system. She begins by explaining how her aunt outlined the intricate processes by which images captured by the eyes are processed and transmitted to the brain's primary visual cortex, located at the back of the head.

"Essentially, light from an image enters your eye and falls on specialized photoreceptors in the retina, called *rods and cones,* which *detect light intensity, color,* and other visual details. The raw visual data is then pre-processed by other retinal cells which, among other things, enhance contrast and detect motion and edges.

"Once the preprocessing is complete, retinal *ganglion cells,* which are bundles of nerve cell bodies, convert the refined signals into action potentials. These electrical impulses travel along the bundled axons of the optic nerve to the *optic chiasma*, a critical junction in the visual pathway, where visual information is reorganized so that each hemisphere of the brain processes input from the opposite visual field. Specifically, the right hemisphere processes visual information from the left field of view, and the left hemisphere processes information from the right field of view."

She points to the diagram and elaborates further on the flow of information.

"From the optic chiasma, the signals travel along the nerve fibers of the optic tract to the lateral geniculate nucleus (LGN). *The LGN acts as a relay station, performing further processing of critical visual aspects such as contrast, motion, and depth perception."*

Finally, Caroline concludes by explaining that after being processed in the LGN, the visual signals are sent to the *primary visual cortex* where the brain conducts a more sophisticated analysis —

identifying features such as edges, orientation, and shapes to create a coherent visual perception of the world around us.

Her parents listen intently, clearly impressed by the complexity and elegance of the visual system.

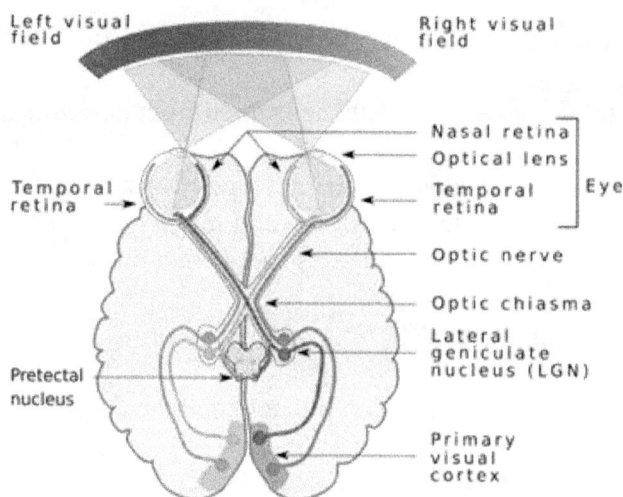

Figure 3. An outline of the pathways from the retina to the primary visual cortex.

Caroline says she learned a lot but found some things totally puzzling.

Mary: "Such as?"

Caroline: "Such as the relationship between visual information and auditory information. For example, if I read a sentence and then hear the same sentence spoken, it means the same thing, but there are two very different pathways involved—the auditory and the visual."

Mary: "Did they explain how that works?"

Caroline: "Not during our first meeting. They wanted to wait until they covered computer systems so they could compare how the brain handles that question versus how machine learning does it. I'll save their explanation for when I tell you about our second meeting, but before that there's something else I'd like to talk about."

Don: "Yes, go ahead," Don encouraged. "Feel free, although I'm not sure how much we'll be able to help."

Caroline: "I'm struck by how quickly information is processed and how fast it flows. Aunt Zoe said the entire chain of events—from detecting a change in the color of a traffic light or the movement of an object, to registering those changes deep in the visual cortex—happens in approximately a hundredth of a second.

"It's not just the speed of the action potential that impresses me—it's easy enough to imagine an electrical signal traveling rapidly. What really amazes me is that the action potential is encoded with information extracted from a sequence of photons arriving from the outside world—and even encoding that information, and without error, takes time."

She paused before continuing.

"Aunt Zoe also pointed out that if you're in a car looking at a traffic light, there's so much more going on than just visual processing—you need to coordinate foot and hand movements, respond to sounds, stay alert for pedestrians, and so on. Different regions of the brain process all this information simultaneously, and they do it very quickly and in a coordinated way that's normally error-free."

Don: "That adds a whole new dimension to the question you raised about the equivalence of auditory and visual information. The brain evolved distinct pathways to process different sensory inputs, store the results in different locations, and communicate seamlessly between those locations."

Mary: "A lot happens all at once, in parallel as they say, as it must if a network with 100 trillion connections is to be fully functional. I'd imagine there's an enormous difference between the way the human brain processes and responds to information, and how computers do it. The brain is essentially a dynamic electrochemical network that processes many kinds of information in parallel including sensory information, memory and motor commands. Computers, on the other hand, process information sequentially—at least that's my understanding. In any case, I'm looking forward to learning how Alice and Aunt Zoe explained the difference between the way information is processed by neural pathways in the brain and how it's processed by a computer."

Don: "Did you discuss other senses, like hearing?"

"No," Caroline admitted. "That crossed my mind briefly, but we didn't have time, and they're both so busy. I didn't want to ask for another meeting."

Mary: "Well, you never know. It's not so unusual for professors to take an interest in mentoring serious students. And you might find auditory processing especially fascinating, given our interest in music—it might even be mutually educational if your father took part in the discussions. A deeper understanding could shed light on the difference between music and noise, or music and speech. In any case, understanding how the brain processes music, as opposed to ordinary language, is a fascinating research question. Maybe someday you'll contribute to that understanding."

Don: "Your mother has a point. The next time you talk to your aunt, you might want to ask her to explain how sound is processed."

Caroline: "Oh, I thought you were going to say, 'Your mother has a point—maybe you'll become a neuroscientist and explain to the world why music brings so much joy.' And who knows? Maybe

I will. But for now, I'll just continue to pick Aunt Zoe's brain and try to understand everything she tells me."

The following Saturday, Caroline met Zoe and Alice for lunch. She ordered a hamburger and a milkshake and, without ceremony, dove right into the discussion.

Artificial Neural Networks Learn by Modulating the Interactions Between Neurons, Just as Biological Networks Do

Caroline: "To what extent do these artificial neural networks simulate the human brain?"

Alice: "They're very different in detail, but they're similar in one important way—each artificial neuron receives input, processes it, and passes the output to other neurons. But before diving too deeply, let's consider how a simple, yet potentially useful, artificial network can be trained to decide whether a cancer will be responsive to a particular therapeutic intervention. In other words, will a particular tumor sample be responsive or non-responsive to a particular therapy given that the sample has certain characteristics? Sound good?"

Caroline: "Sounds good."

Alice: "The algorithm, of course, needs to be trained on data. I'll use artificial data to explain the ideas, but once you understand the concepts well enough, you might want to experiment with a neural network using actual data. Let's say we have 400 samples collected from patients—200 of which are responsive to the therapy and 200 that are not. Let's also suppose scientists have identified five proteins that, on average, are at higher levels in the responsive samples than in the non-responsive samples. The goal is to train a network that can correctly classify a tumor with unknown responsiveness, even though the categories (responsive/non-responsive) in the training data aren't cleanly separated by the concentrations of the five proteins."

Caroline: "This assumes that the protein levels in the sample you intend to test have already been measured, right?"

Zoe: "Yes, you would know the protein levels but not whether the tumor is responsive to treatment."

Alice hesitated for a moment, realizing that she might be moving too quickly. She turned to Caroline and said, "The example I have in mind is going to introduce concepts like *high-dimensional spaces*, which are likely unfamiliar to you. So, before we dive into it, we'll need to review some basic geometry."

Caroline: "High-dimensional space? You're right—I have no idea what that even means. It sounds like something out of science fiction."

Alice laughed. "I thought the term 'high-dimensional space' might grab your attention. When I first learned about spaces with more than three dimensions, I was baffled too. It sounded like science fiction to me as well. But believe me, it's actually almost prosaic once you break it down."

Alice continued, "First, let's define *a vector. It's simply a set of rational numbers, but it can also be interpreted geometrically."*

Caroline: "What's a *rational number?*"

Alice: "It's a number that *can be expressed as the ratio of two integers*, like 3/7, -9/2, or 17 (which can also be written as 34/2).

"I'm sure you remember that in trigonometry, rational numbers were represented on a straight line. If you designate any point on the line as zero, positive numbers increase as you move to the right, and negative numbers increase in absolute value as you move to the left.

"Now, to define a two-dimensional plane, we draw a second line perpendicular to the first one, passing through the zero point. On this new line, positive numbers increase as you move inward and negative numbers increase as you move outward. These two lines are usually referred to as the x- and y-axes. The point where they intersect, x = 0 and y = 0, is called the origin."

Alice draws a figure to illustrate.

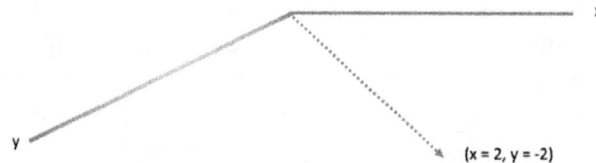

Figure 4. A portion of a two dimensional plane: x increases from left to right, y increases into the page. The axes can in principle range over all rational numbers, from -∞ to +∞. The figure shows only one of the four possible quadrants, where x ranges from 0 to ∞, and y ranges from 0 to -∞.. As the figure indicates vectors in 2 dimensions can be represented as an arrow or as a set of numbers.

"A three dimensional space can be defined by adding a third axis, an up-down dimension, which is usually labeled z, perpendicular to the other two."

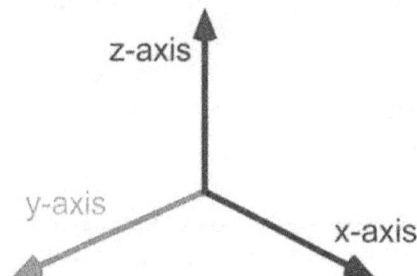

Figure 5. A three dimensional vector—the coordinates of a point in a 3 dimensional space, represented by an arrow whose head is at that point. The diagram shows only one of the (2 x 2 x 2= 8) possible octants.

Alice again draws a picture, labeling as an example the point x= 3, y = -2, z = 6 which she explains is usually written (3, -2, 6) for simplicity. She tells Caroline that vectors can be added and subtracted.

Alice: "Each position in the vector is called an element. *Vectors are added by summing their corresponding elements*; for example, the vector (3, -2, 6) when added to the vector (4, 1, 9) gives the vector (7, -1, 15)."

Caroline: "What happens in 4 dimensions—how do you draw a 4D arrow?"

Alice: "You don't."

Then Alice adds the amazing mathematical punchline: "*Although dimensions greater than 3 are difficult to visualize, the vectors in any number of dimensions can nevertheless be added and subtracted just as in the spaces that you can visualize, simply by adding or subtracting their corresponding elements.*"

Caroline: "And no matter what size the space, no matter how many dimensions, you can always add and subtract, just like you do with single numbers?"

Alice: "Yep."

Caroline: "What about multiplication and division?"

Alice: "You would ask that! Multiplication is a little more complicated, but it's not necessary to delve into that right now. And division—it's not possible."

Caroline: "What? *You can add, subtract and multiply, but can't divide?*"

Alice: "That's right," Alice says, changing the pitch of her voice between the two words, "and *there's more than one way to multiply vectors.*"

Caroline: "That's mind-boggling."

Alice: "Yes, it's fascinating stuff. You might want to try reading about it, but right now we're almost ready to get back to the example."

Caroline: "We're still not ready?"

Alice: "Not quite. To have any idea of what might be going on in higher dimensions, it's best to start by considering a 2-dimensional version. So let's suppose there are only 2 proteins rather than 5, and to keep the example really simple, let's say we have only 40 samples—20 tumors that respond to therapy and 20 that don't."

Caroline: "I hate to be a nuisance and to ask for information that might be largely peripheral, but it would help me if you could be a little bit more concrete about the proteins, especially since I might want to do some research on my own."

Zoe: "OK, I'll pick proteins called MCT1 and telomerase. These happen to be involved in skin cancer—and I'll leave it to you to look up their functions if you're interested. It's easy to do, and, as you correctly noted, the functions of these proteins are largely peripheral to our discussion."

Alice: "Getting back to our example, you can visualize the concentrations of 40 pairs of proteins as being points in a 2-dimensional plane, with the Y-axis labeling the concentration of one of the proteins and the X-axis labeling the concentration of the other. The goal is to find a line that completely separates the responsive samples from the non-responsive samples. It turns out that, in this case, that can't be done using a straight line, or as they say in the business, the data are not linearly separable."

Alice retrieves a plot from her computer that illustrates what she said, but just as she starts commenting, Caroline interrupts.

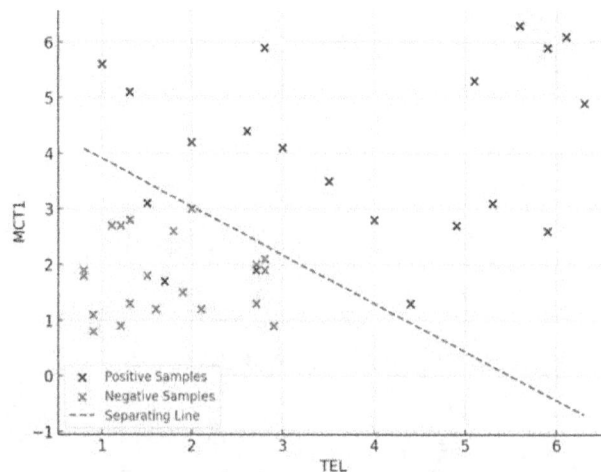

Figure 6. The dashed line uses a standard algorithm to separate tumor from normal samples based on two proteins whose concentrations are, on average, higher in the former category. Neither this method nor a neural net achieves perfect separation.

Caroline: "Why can't you separate them in 2 dimensions by using a squiggly line?"

Alice: "Let's back up a bit and recall something very fundamental about polynomials. A straight line has two parameters—its slope and intercept. A squiggly line, depending on what you mean by "squiggly," will have multiple parameters. I'm sure you already know that a first order polynomial, i.e. a polynomial of degree 1, will have 3 parameters (ax + by + c = 0, which reduces to a two parameter polynomial $y = \alpha x + \beta$ if $b \neq 0$, where $\alpha \equiv -a/b$ and $\beta \equiv -c/b$). If you use higher degree polynomials, the number of parameters grows large quickly, or combinatorially as the mathematicians say. A second degree, or quadratic, polynomial has 6 parameters, a third degree polynomial has 10 and in general polynomial of degree k has (k+1) (k+2)/2 parameters."

Caroline is listening intently, and though she's not saying much, she's diligently taking notes.

Alice: "We don't have time right now to go into the problem of overfitting—that is, using more parameters than is justified by the amount of available data—but the key point to remember is that, in general, a separator won't be useful if the number of free parameters is comparable to, or greater than, the number of data points in the smaller of the two categories (20 in this case), and ideally, it should be much less. *A famous mathematician, John von Neumann, once said that he could fit an elephant with a four-parameter curve, and with five parameters, he could make its trunk wiggle.*"

Alice continues: "The difficulty is that *a separator with too many free parameters is not generalizable—it will fail when presented with new data.* So, the answer to your question is, yes, you could try a quadratic or even a cubic function and might get separation, but since the likelihood of the separator being generalizable decreases rapidly as the number of parameters becomes comparable to the number of data points, it's unlikely to be useful."

At that point, Alice abruptly shifts the discourse.

Alice: "Since two proteins are not sufficient to separate the samples, let's try three, and this time, let's use a neural net."

She then shows Caroline a diagram (Figure 7) consisting of four rows: three circles at the bottom, five circles in the next two rows, and a circle at the top—with successive rows connected by arrows. She refers to the diagram as a graph, the circles as nodes, and the lines between them as links, explaining that some links were omitted for clarity.

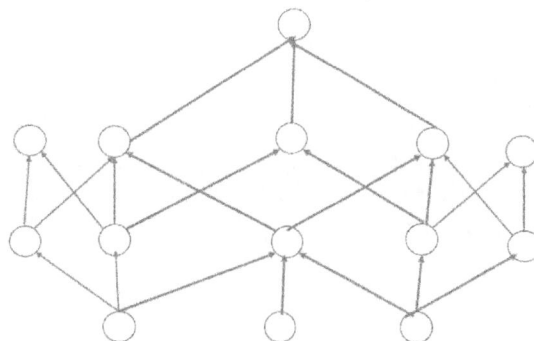

Figure 7. A network with a 5,5,1 architecture. The input layer is at the bottom, and the three rows above it are processing layers.

Alice: "The three nodes in the bottom row represent the input—the concentrations of the three proteins for each sample presented to the network. The next two rows represent neurons. Each neuron can be in one of two states: active (represented by 1) or inactive (represented by 0). The links represent the flow of information across a synaptic cleft via neurotransmitters, and crucially, not all links have the same strength. In other words, the information flow depends on both the origin and destination of the signal. The graph captures the details of this flow by assigning different weights—just numbers—to different links.

223

"A graph with nodes and links arranged in this pattern is said to have a 5-5-1 architecture: each neuron in the second row receives input from every sample, and those neurons that fire send a signal to every neuron in the third row. Finally, the neurons in the third row that fire send a signal to the output neuron at the top, whose state depends on the flow of information in the network."

Caroline: "What determines whether a neuron is active or inactive? I mean, how does it decide whether or not it fires?"

Alice: "That's a good question. The answer is the sum of the weights feeding that neuron: if the sum exceeds a threshold, the neuron fires; otherwise, it's quiescent. Now here's the crucial point: since we know which samples are positive and which are negative, we know what the value of the output neuron must be for every sample—but we don't know the weights and thresholds that will give the correct output."

Alice then stops her explanation and asks Caroline to calculate the number of parameters, that is, the number of weights and thresholds required to characterize the network.

Caroline is silent for a moment, then starts thinking out loud.

"Let's see, there are three inputs that signal each neuron in the first layer, so I guess each neuron is characterized by four parameters: a weight from each of the inputs and a threshold that must be exceeded for the neuron to fire. Since there are five neurons in the first layer, that makes 20 parameters. For the second layer, the number of parameters is $(5+1)\times 5 = 30$, and since the output neuron requires another six parameters, the total is 56 parameters."

Caroline's a bit surprised.

"Isn't that too many parameters with the amount of data we have?"

Alice: "It certainly is. How much data do you think we should use?"

Caroline: "Would 120 positives and 120 negatives be safe?"

Alice: "It might be—let's give it a try."

But before going further, Zoe asks Caroline whether she has any thoughts on how all those parameters might be determined so that the network correctly responds to every piece of data in the training set.

Caroline is stunned and says she hasn't the slightest idea.

Then after a moment's reflection she says: "I suppose the next step is to find the weights and thresholds that give the correct output for any of the samples in the training set, but I have no idea how that can be done."

Alice: "OK, I'm sure this will be entirely new to you, and it will also be just an outline, but here's how it's done. The first step in training the algorithm is to assign a randomly chosen weight to each parameter."

Caroline: "I'm already in the dark. Can the parameters really be anything?"

Alice: "That's a good question—in principle they can be anything, but it's best to keep them small."

Alice continues, explaining that the goal is to find values for these 56 parameters such that the network will classify the samples with a small enough error to make the network useful.

Caroline: "I'm not sure what that means."

Zoe: "It means, ideally, finding a network for which the number of misclassified positive and negative samples is 0."

Alice: "It's easy to develop a network that performs perfectly during training, but it's less easy to find one that performs with no error during testing—and almost none of them perform perfectly when applied to entirely new situations.

"Getting back to what I was saying, for each sample, the output during training is known, either 1 or 0—so if the network were performing perfectly, the sum of the outputs for the positive and negative samples would be +120 and -120, respectively. But when training begins, those will not be the computed outputs; in other words, there will be an error. The goal is to reduce the square of that error to 0 by changing the weights and recomputing the sum of the outputs."

Caroline: "Why the square of the sum and not just the sum itself?"

Alice: "I think you can answer that question yourself just by imagining a series of responses that are largely incorrect but sum to 0 or near 0 because the pluses and minuses cancel."

Caroline: "OK, I think I understand, but I still have a question—are the weights just changed randomly?"

Alice: "Great question—they're not just chosen randomly. Choosing weights during each iteration so that the network arrives at a specified degree of accuracy as quickly as possible is a problem in optimization, which is an active subspecialty of computational science. You're probably three or four years away from being able to delve into it, so it's best to set that discussion aside. For now, the bottom line is that numerous approaches can be taken to make the search more efficient than just randomly choosing a different set of weights for each iteration. Hundreds of iterations might be required to bring the error down to 0, but eventually, that goal will be met. Error reduction is followed by testing, and if the network meets pre-specified characteristics, whatever they might be, it is applied to make actual decisions."

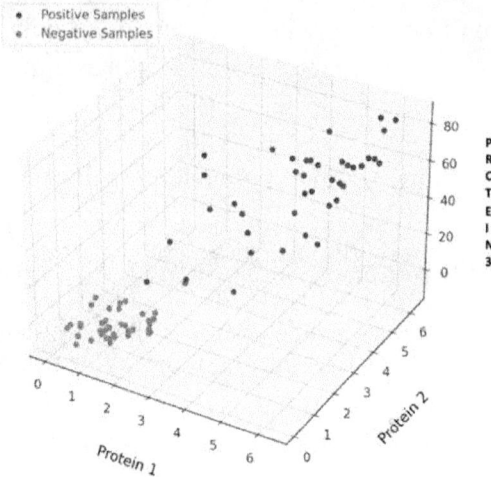

Figure 8.. An example of perfect separation in three dimensions which could not be achieved in 2 dimensions, based on three proteins whose concentrations are on average higher in tumor samples than in normal samples.

Zoe: "This simple model that Alice just discussed has a couple of important takeaways. One is that the nodes capture some of the essential features of biological neurons. As you'll recall, they too respond in an all-or-none fashion: if the total input to a neuron exceeds a threshold, an action potential propagates along the axon and causes the release of neurotransmitters. These neurotransmitters then diffuse across the synaptic cleft and bind to dendritic receptors on the postsynaptic membrane of neighboring cells, thereby contributing to their inputs. Another takeaway is that the model is not so different from those sometimes used in research by cancer geneticists."

Caroline: "Are there simpler neural nets that also give perfect separation and are error-free when tested on another set of samples?"

Alice: "That's a great question. In fact, although I didn't try very hard, I found one with a 3,3,1 architecture that also performed perfectly during training and testing."

Caroline: "Is there a way to take an informed guess at which might be better before applying them to samples that have never been classified?"

Alice: "Yes, a useful piece of information is the average distance of the data from the decision boundary, for example, the sum of the squares of the distances of each data point from the decision boundary. For the 5,5,1 architecture, the result is 19.3."

Caroline: "19.3? What are the units?"

Alice: "The square of the concentration, in whatever concentration units are being used.

"For the 3,3,1 architecture, the result is 11.87."

Caroline: "How about the original method? I think you referred to it as regression. Did you try that in three dimensions with the current set of samples?"

Zoe: "Wow, you really push examples to the limit!"

Alice: "Yes, she does, and it was another great question. The answer is, I haven't tried it, but why don't we do it right now?"

Alice runs the program and finds that a linear regression model does not separate the training data perfectly, although it does a good job, achieving an accuracy of 97.5%. She explains that even though the percentage is high, the performance is really unsatisfactory because the best chance of obtaining perfect separation is with the training data. She also points out that the sum of the squares of the distances from the hyperplane of the correctly categorized data is 2.2. She then asks Caroline if she can guess the significance of those distances.

Caroline: "I'm not sure I understand, or maybe I just can't imagine."

Zoe, addressing Caroline: "Remember, these are hypothetical error-free data—but suppose they were real data with errors. What might happen to samples that were correctly classified but were near the hyperplane?"

Caroline: "Oh, I think I get it. If the data are imperfect, the chance of misclassification is greater when the sum of the squares is small than when the sum of the squares is large."

Alice: "Yes, very good. What we would say in the business is that, of the three architectures, the 5,5,1 is the most robust against error."

Caroline: "That was very helpful. Now I'm ready to take a deep breath and try to learn something about the monster neural nets used to model languages."

Large Language Models (LLMs) Are Neural Nets That Can Respond to Long, Complex Queries With Proper Syntax and Semantics in Many Domains of Human Knowledge

Alice: "There are many differences between the artificial neural net we just discussed and the neural nets used in large language models, the most obvious being that an LLM has hundreds of billions of parameters. But there are also important differences in the actual algorithms, which are too complicated to delve into."

Caroline: "The difference in size is incredible, but I'm far from clear on how a machine learning algorithm like GPT-4 works."

Alice: "Wait—I've yet to begin. But before I do, it's important to recognize that the best way to understand these systems—or at least develop an intuitive feel for their capabilities and limits—is to play with them a bit. Keep in mind: we're entering the most transformative era in human history. To meet the many unknown challenges ahead, it's crucial to be as prepared as possible. And if history is any guide, those challenges will arrive sooner than most people expect."

Caroline, not surprisingly, is eager to follow up on Alice's suggestion. But there's a flicker of unease on her face—perhaps because she's realizing, for the first time, how rocky and tortuous the road ahead might be. She admits she has no idea how to even begin "playing" with these systems.

Alice: "One way to start is by learning how to query the algorithms effectively—how to ask questions in a way that gets you accurate, meaningful responses. My schedule is flexible late Thursday afternoon. I'd be happy to meet with you then and help you get started."

Caroline: "Oh, that would be great. I'd really appreciate it."

Alice: "It would be my pleasure. But even before we meet, you might want to think about the kinds of questions you'd pursue if you had the right training."

Caroline: "I'm afraid I'm at a loss."

Alice: "I'd be shocked if you weren't. But just to get you going: one of the key topics we're exploring is how large language models internalize human values—things like honesty, privacy, liberalism, and so on. If I were a high school student, I'd start reading, thinking about what I read, talking to people in the field, and querying these systems on as many topics as I could."

Caroline: "OK, thanks."

Alice: "You're welcome. For now, I'll try to give you a sense—just a rough outline—of how some of these algorithms work."

Caroline: "My seatbelt is buckled."

Alice: "Here we go. And just so you know—by the time I'm done, you still won't understand exactly how something like GPT works. But you'll have a decent 5,000-foot view.

"Let's start with this: a computer works with numbers. To process text, it first converts each word—or word fragment, called a *token, which on average is approximately 3/4 of a word*—into a vector with hundreds, and for some models even thousands, of dimensions. This is called *embedding*."

Caroline: "I'm baffled. I can't imagine how a single word, let alone a word fragment, can have hundreds of characteristics."

Alice: "I know it's difficult to imagine. Each element is not a simple word; it's a combination of features learned during training. For example, different elements might indicate positive, negative, or neutral feelings; others might indicate how strong the feelings are, and so on. It's abstract, and I don't think anybody fully understands it. The bottom line, however, is that each element of the vector represents some nuanced shade of meaning or feeling that the algorithm has learned from examples of that token in an extremely large number of different contexts."

Caroline responds, saying she has a vague understanding and then tests herself by asking whether *the word bank as in 'you can bank on it' and the word bank as in 'my mother works for a bank' have different vectors because the surrounding words change their meaning*?"

Alice: "Yes, that's the general idea. But just to be clear, the embedding for bank is initially fixed, but the **final vector representation** of that token *within a sentence* is ***contextualized*** by the deeper, transformer layers of the network."

Caroline: "So, a series of words is represented by a series of multidimensional vectors, or tokens, and the particular element of the token that is used depends on all the surrounding tokens. How many tokens do LLMs have?"

Alice: "About 50,000, representing approximately 33,000–38,000 English words."

Caroline: "Got it. Now, what about the neural net itself? You said there were upwards of half a trillion parameters, but you didn't say how many layers a typical LLM has."

Alice: "There's no fixed relationship between the number of layers and the number of parameters. An LLM can have tens or even hundreds of layers. We'll have to leave it at that because it's a complex subject, and there's very little time left to complete this discussion."

Caroline: "OK, at some point I'd like to learn more, but I'm even more curious about the amount of data required to determine so many parameters—and where it comes from?"

Alice: "Models like ChatGPT are trained on trillions of words and billions of examples derived from huge data warehouses of publicly available texts, including journals, books, and websites. During training, it's given pieces of text, called prompts, and it learns to predict the next token."

Caroline: "How long is a prompt?"

Alice: "The context windows of the latest LLMs can handle hundreds of thousands of tokens, and the number is increasing rapidly."

Caroline: "And the output?"

Alice: "For any set of weights—and by weights, I mean the strength of the connections—the output will be a probability for each token in the vocabulary. So if the vocabulary has 50,000 tokens, the output will include 50,000 probabilities, one for each token."

Caroline: "Is the token with the highest probability chosen?"

Alice: "Often, but not always. Sometimes randomness is added to make responses less repetitive or more creative. After choosing a token, the model updates the sequence and repeats the process until the response is complete.

"During training, the output is known, and initially, the response to a prompt will be incorrect. The difference between the known output and the computed output can be calculated. The goal is to reduce the sum of all those differences over all prompts to an acceptably low level, and that's done by adjusting the weights. Learning means adjusting the weights until the network responds with syntax, semantics, and information appropriate to the query. However, the error is rarely reduced to zero: such accuracy is not always achievable, nor is it desirable, both because of excessive cost and the possibility of overfitting."

Caroline: "It's a bit confusing."

Alice: "I know, understanding how the error is reduced requires graduate-level mathematics. Your uncle understands it, and maybe after you've had several years of college mathematics, he can explain it to you. But for now, you get the rough idea, don't you?"

Caroline: "I guess so. Somehow there's a way to adjust the weights so that the error is reduced to a very low level."

Alice: "Yes, and once that happens, the weights are fixed, and the network will generate human-like responses to any query from any field of knowledge. But remember, although progress is astoundingly rapid, it's still not human thought, and the responses are sometimes inaccurate and biased. It's more about processing information in a way that seems intelligent.

Nevertheless AI, especially when combined with advances in robotics, has the potential to transform virtually every aspect of human life—science, the arts, entertainment and finance, and just about any service industry you can think of."

Caroline: "Alice, you're beginning to sound like my father—using words I don't fully understand. I know what science and art are. And Mom's in finance, so I have an idea about that—but what's a service industry?"

Alice: "I guess you haven't had an economics course yet."

Caroline: "I'll be taking economics next semester."

Alice: "The global economy, for simplicity, is sometimes divided into three sectors. The relationship between the first two is direct: raw materials are provided to the manufacturing sector, which converts them into tangible products. The service sector provides intangible goods, such as advice to businesses and individuals. For example, most professionals—such as accountants, physicians, lawyers, and teachers—are in the service sector."

Caroline: "OK, I get it, but can you give me a specific example? For instance, how will AI impact medical services?"

Alice: "Imagine a physician who has seen 100 million patients, knows everything ever written about diseases and their treatments, and knows more about you than you or any group of human physicians do—from your medical history to your genetics."

Caroline: "Wow, that would be some physician."

Alice: "Yep, and AI physicians may well be upon us sooner than most people realize. What's missing is rigorous and extensive testing—to demonstrate that when there are disagreements between AI and human experts, the AI is almost always correct. I expect that we'll see very convincing evidence of that over the next decade or so."

Caroline: "What happens then?"

Alice: "Who knows? It's going to be disruptive, and we as individuals and as a society need to start having spirited discussions immediately if there's to be any hope for a reasonably smooth transition. But that's a subject for another day."

Epilogue

A decade had passed since Caroline's senior year of high school and not surprisingly it was a time marked by both triumphs and challenges. Caroline had grown very close to Zoe, and she had also developed a warm rapport with Zoe's brilliant fiancé, Barry Bruce. Barry's encouragement and kindness had a deep impact on Caroline during those formative years, yet as life often reminds us, destiny sometimes differs from the expected.

An inflection point occurred when Barry faced a pivotal decision that weighed heavily on both him and Zoe. Driven by an unwavering ambition to reshape the scientific enterprise, he accepted the position of director general at a new European Union institute. The prospect was thrilling yet daunting, as it meant leaving behind the life they'd started to build together. Although they initially planned to marry, the strain of distance and their diverging career paths gradually dimmed the glow of their shared dreams. At around the same time, Zoe's career flourished, culminating in a full professorship at the University of Chicago, where cutting-edge facilities allowed her to tackle groundbreaking research.

Meanwhile, Caroline's friendship with Noreen began to change. Their interests and lifestyles were moving in opposite directions, and after they entered college, they slowly drifted apart. Noreen was awarded a full scholarship to Harvard and graduated summa cum laude before earning a PhD in computer science at MIT in record time. Now regarded as one of the nation's brightest and most promising young scientists, she lives in Cambridge with her partner, Raquel.

Caroline stayed in New York and entered NYU as a mathematics major. She had heard, well before she graduated from high school, about NYU's legendary Courant Institute and its standing at the pinnacle of applied mathematics. The combination of her love for the City and NYU's reputation was enough to incentivize her to apply for early decision—and perhaps she had an additional motive: she had a steady boyfriend, Sam—the boy down the street who was a shortwave radio expert and engineer to be.

Caroline's relation with Sam, however, was destined to unravel. The late-night conversations grew sparse, and shared laughter became a rare commodity. Caroline felt a subtle distance creeping in—a silence that words couldn't fill. In just over a year, they parted ways. It was a quiet ending to a once-vibrant chapter, leaving Caroline with a mix of melancholy and a newfound sense of self-reliance.

In any event she was accepted by NYU, and the following fall she entered the College of Arts and Sciences, naively expecting to spend four years taking almost all math and science courses. As it turned out, the requirements for a Bachelor of Science degree included one semester of economics, and by the time she had completed the required course she had undergone a true epiphany.

The subject matter was unlike anything she had previously encountered, and she absolutely loved it. Her newfound captivation resulted in part from the realization that economics is very mathematical, and her mathematics background enabled her to dive deeply into subjects such as game and decision theory, and various methods that employ mathematical statistics. In her senior year she didn't hesitate to apply to the famed PhD program in economics at the University of Chicago, and the following fall found her living not too far from her aunt.

Caroline was in her element: the curriculum was challenging but rewarding, her fellow students were friendly, and the Economics department faculty, which included seven Nobel laureates, were remarkably approachable. But most rewarding of all were the several hours every other weekend spent with her young cousin, occasionally attempting lightheartedly to explain their shared DNA. Yes, Zack Angstrom and Zoe DiStefano, the couple who introduced her parents to one another, finally closed the loop, and the backstory was hardly one of coincidence.

Don Angstrom of course knew that his sister-in-law was negotiating with the University of Chicago, and since he communicated regularly with his brother, Zack also knew.

Zack enjoyed the mountains of Northern New Mexico but nevertheless missed the museums, symphonies and the gourmet restaurants of great cities, and he also missed the intellectual breadth of a university, so he occasionally considered moving. It therefore came as no surprise to Don that as soon as Zoe accepted the position, Zack contacted her, and the next day he contacted colleagues at Argonne National Laboratory. Six months later he had begun his new position as a senior scientist at Argonne and adjunct professor of physics at the University. He had arrived in Chicago at almost exactly the same time as Zoe. Two months later they were married and ten months after that Caroline's double first cousin, Mario, was born.

Mary and Don, still living in Manhattan, missed their regular get-togethers with Zoe, but they had a wide circle of friends and colleagues and engaging professional lives. Don's reputation as a virtuoso pianist grew over the years and he was now a member of the pool of piano players employed by the New York Philharmonic, and asked to perform from time to time depending upon the repertoire.

Meanwhile, years earlier when Caroline was in her senior year of high school, her mother made a bold decision to leave Packard Analytics. Fueled by determination and a vision for the future, she founded her own hedge fund. The venture surpassed her wildest expectations, its success enabling her to establish a Family Office.

But it wasn't just about personal achievement; Mary wanted to make a difference. She founded the Mary and Caroline Angstrom Foundation, a testament to her commitment to using her success as a force for good. The purpose of the Foundation was to help support small farmers in sub-Saharan Africa and the Asian Pacific who found themselves in serious economic straits because of climate change. She used most of her earnings to establish an endowment and aimed to eventually donate some 95% of her wealth to the Foundation.

Although Don had had a very full professional life, he nevertheless found time to sit on the Board of the Foundation and to visit countries in sub-Saharan Africa on its behalf several times a year. During his travels, he formed deep connections with communities across southern and western Africa. He became captivated by the vibrant rhythms of the Ashanti drummers in Ghana, whose beats seemed to echo the very heartbeat of the land. In South Africa, the soulful chants of the Zulus resonated with him, their music a tapestry of history and resilience. The lively sway of the Congolese rumba from the Democratic Republic of Congo drew him into its infectious groove. Immersed in these rich musical traditions, Don didn't just listen—he participated, learning to play the djembe and the marimba, instruments that bridged cultures and brought people together in joyous harmony.

Mary often remarked with amazement, "Don, I don't know how you find the time to breathe." Yet, despite his whirlwind schedule, Don always made time for the things that mattered most, including their shared passion for making a positive impact. This mutual dedication only strengthened their bond over the years.

A great deal happened during the past decade, and although some things fell apart other relationships tightened and strengthened. But there was one striking, though perhaps minor, loose end: Don and Caroline never did get together on those rainy Saturdays to address some of the deep questions that were unanswered in their casual discussions. The reason perhaps was that Caroline's knowledge and understanding of science quickly outgrew that of her parents, but she also had a strong desire to achieve intellectual independence—to search for answers on her own. A striking example of that drive was the collection of books described below, which were assembled and read during the summer before she entered NYU.

Finally and most importantly, as Caroline grew older her perspective on her mother's life and choices transformed markedly. What she once dismissed as an almost insatiable love for money now appeared as a powerful drive to create meaningful change. She admired the way Mary channeled her wealth into helping struggling farmers half a world away, realizing that her mother's ambitions had always been about building something greater than herself.

The breadth of understanding that grows with maturity had at last replaced the know-it-all attitude of youth, leaving a deep sense of gratitude for the values her mother had instilled in her. She was an adult and on her way to becoming a leading economist— but perhaps somewhat ironically the accumulation of wealth was not one of her goals. In fact her main aspiration echoed the timeless wisdom of Aristotle's Metaphysics: "All men by nature desire to know." But in Caroline's case the pursuit of knowledge wasn't just academic—it was deeply personal. She was driven to not only uncover economic insights but also to share them, teaching others and striving to shape policies that could improve the lives of countless people. Her journey was about more than material success; it was about making a meaningful difference.

NB: Caroline is loosely based on Zoe Huang, the daughter of Xin Huang and Zhiping Weng, and an undergraduate at the University of Chicago. A number of Zoe's friends read *Curious Caroline* and were impressed that Caroline was in part based on Zoe. Then one of them, a computer science major—in fact it was Alice Turing's daughter— had a brainstorm. Wouldn't it be interesting if

Caroline and Zoe could talk to one another. In fact wouldn't it be interesting if we could all talk to Caroline. Well, the idea spread like an epidemic—as though it were a highly contagious virus. There was no scarcity of students eager to create a Caroline AI, and no scarcity of faculty willing to guide and advise. Although data were limited, the requirements were not nearly so great as they would be for creating a general purpose AI, and beyond that they knew that as people spoke with Caroline—and there'd be plenty who wanted to, training data would rapidly accumulate and her responses would become quicker and more natural.

Faculty as well as students soon became engaged in the project, and a number of them suggested that Caroline's dataset be augmented with all information that an intelligent high school student could know, and that it be done in a way that preserved Caroline's distinctive thought patterns. The result would be a supersmart high school student—in fact the smartest in the world—with Caroline's personality.

Then two members of the team, one from Physics and one from Electrical Engineering, suggested something even more radical: that Caroline be represented by a talking hologram. They pointed out that sound could be added to holograms, just as it can be added to film, though the technology is somewhat different. Some members of the team are even talking about bringing the entire Angstrom family to 'life', and augmenting them with an abundance of recent knowledge. They're dreaming big, eagerly awaiting the day when they'll hear Don perform, Mary sing and Caroline play her violin. As far as anyone knows this will be the first time characters in a story will be effectively brought to life—not organic life, but silicon based and coming pretty close to being human with nuanced personalities created by a storybook author.

As to be expected, the project is, as it should be, causing a firestorm of controversy, with ethicists and other concerned faculty from all components of the University out in full force, reminding the enthusiasts about the monumental known and unknown risks. Nevertheless, the project is underway, and it seems Pandora is already out of the box. What happens next? Chi lo sa, as Aunt Zoe might say."

Author's Note: Everything in the above note (NB), after the first sentence, is purely fictional. It will be a while before a real person will be able to converse with Caroline, maybe as long as a year.

Caroline's Notebook

Conversation 1

I just learned a lot of confusing and, in my opinion, not especially important facts. Well, maybe they're interesting to a historian or a sociologist because they say something about the changing sociology of science. It appears that scientists communicated poorly about standards and the definitions of units, which are no small matter for the development of quantitative sciences. The situation with units is so problematical that even Richard Feynman, who Dad called one of the most brilliant theoretical physicists of his generation, said that "For those who want some proof that physicists are human, the proof is in the idiocy of all the different units which they use for measuring energy".

I was particularly struck by the fact there are several different systems of units. The US uses so-called Imperial units which were introduced by the British. The Imperial unit of mass is the pound, the unit of force is the pound-force, and the unit of length is the foot — and of course there are related units such as inches, yards and miles for length, which are needed to avoid numbers that would otherwise be unwieldy. Part of the reason mass and weight are sometimes confused is that the pound is commonly used for force as well as for mass, and they're differentiated only when necessary by using pound-force (lbf) and pound mass (lbm). By definition 1 lbf accelerates 1 lbm at a rate of 1 foot/sec^2 (1 ft/s^2). In the metric system (SI units) the corresponding units for mass, weight and distance are the kilogram (kg); kilogram-force (kgf), the meter (m) — and of course additional terms for their multiples and fractions, such as kilometer (km) and centimeter (cm) for distance. However, kgf is no longer used — instead the SI unit is called the newton (N), which is the force required to accelerate a mass of one kg at a rate of 1 meter/sec^2 (1 m/s^2).

When I weigh myself, the scale must measure a force; namely the force between me and the Earth, but it displays mass, i.e. the scale divides the force it measures by the gravitational field intensity, g. In imperial units that would mean the measured weight is divided by 32.174 to obtain the number that's displayed. For me the scale shows 110 lbs, which is my mass although it's called my weight. Wow, what a mess! If I were to weigh myself in Europe the scale would display a mass of 50 kg, whereas it would measure the force between me and the Earth, which is 50g = 50 x 9.8 = 490 N.

But underneath all of this is an awesome discovery: Newton's law of gravity which applies here on Earth and as far as we know everywhere in the Universe. It says that the force, F, between masses m and M whose centers of mass are separated by a distance r is

$$F = GmM/r^2 \equiv gm$$

where G, the universal gravitational constant, for the planet Earth is approximately 6.674 x 10^{-11} N m^2 kg^{-2}.

What would I 'weigh' on the Moon? I'm putting weigh in quotes because scales display mass. Consequently the scale would read 50 kilograms provided the Moon's g (1.63 m/s^2) is used as a calibration factor. My weight, however, i.e., the force between me and the Moon would be about 1/6 of what it is on the Earth.

The gravity defying fluid

I struggled off and on for more than a week, then I finally had a breakthrough, which in retrospect seems very obvious. The loss of fluid causes the water pressure to change from $W = \rho_w g d \pi r^2 / \pi r^2 = \rho_w g d$

to $W' = \rho_w g (d - \delta) = W(1 - \delta/d)$

Then I realized that when fluid leaves the column, there's more room for air molecules, so that the pressure exerted by the air column also changes, and in particular since the length of the air column increases from h to h $+\delta$ the air pressure drops to $P_{atm}/(1 + \delta/h)$.

I'll call P' the total internal pressure at the base of the straw, which is a fraction of the initial atmospheric pressure. And I know that δ is directly related to the amount of fluid that leaves the straw. Specifically, the fractional loss is $\rho_w (d - \delta) \pi r^2 / \rho_w d \pi r^2 = (d - \delta)/d$.

Then

$$P' = P_{atm}/(1 + \delta/h) + W(1 - \delta/d)$$

Since at equilibrium P' equals the atmospheric pressure, P_{atm}

$$P_{atm} = P_{atm}/(1 + \delta/h) + W(1 - \delta/d).$$

I made the equation look a little neater and also reduced the number of symbols by defining the dimensionless quantity

$$\alpha \equiv W/P_{atm} = 588/100{,}000 = 0.0059.$$

Here I estimated the pressure $W = \rho_w g d = 588$ Pa by assuming the straw is buried to a depth of 6 cm.

When I substituted the expression for α into the equation for P_{atm} I obtained a quadratic equation for δ.

$$\delta^2 - \delta[\frac{d}{\alpha} + h - d] - hd = 0$$

I again made the equation look neater defining

$$B \equiv [\frac{d}{\alpha} + h - d]$$

Then

$$\delta^2 - B\delta - hd = 0$$

which has the solution

$$\delta = -B/2 \left(1 - [\sqrt{1 + 4hd/B^2})\right)$$

I was so excited by this result that I showed it to Noreen who is the best student in my class, and she almost immediately recognized that the expression can be simplified. She explained that it's always important to have an idea of the numerical values of the quantities in an equation, So we estimated the value of $4hd/B^2$ and found that it's much less than one. Noreen then showed me a very interesting relationship:

$\sqrt{1 + x} \sim 1 + x/2$ when x <<1. I then estimated the accuracy of the radical for different small numbers and verified that it's applicable to the quadratic equation I derived, and finally ended up with a very simple and accurate expression for δ.

Conversation 2

Wave interference

I know that if $A_1 = \alpha \sin \theta$ and $A_2 = \alpha \sin \theta$ then their sum is

$A_T = A_1 + A_2 = 2\alpha \sin \theta$. For a physical wave such as light or sound $\theta = \omega t$ where ω

is 2π times the frequency of oscillation.

In this simple situation with coinciding peaks and equal wavelengths, the waves are in phase; their peaks and troughs coincide, but I wonder what happens if I add 2 waves that are out of phase. The sum would be

$$A = \alpha \left[\sin(\theta + \phi) + \sin \theta\right]$$

where ϕ is the phase difference, the number of radians (or degrees) by which one wave leads or lags the other. An easy way to write this as a single term would be to use an identity that we recently learned in trigonometry, which relates the sum of two trigonometric functions to a single function.

sin(a) + sin(b) = 2 sin((a + b) / 2) cos((a - b) / 2)

It's just a matter of substituting a = $\theta + \phi$ and b = θ to obtain

$$A = 2\alpha \cos(\phi/2) \sin(\theta + \phi/2)$$

For waves travelling in opposite directions

A(x, t) = $\alpha\sin[2\pi f(t + x/v)]$ - $\alpha\sin[2\pi f(t - x/v)]$

Now a = $2\pi f(t + x/v)]$ and b = $2\pi f(t - x/v)$ Therefore

$$A(x, t) = 2\,\alpha\cos(2\pi ft)\,\sin(2\pi fx/v)$$

Now what? Dad said the slope can be found using simple calculus. If it's really easy I'll log in to a Kahn Academy lecture and see how far I get.

A week has gone by and I watched some elementary lectures and learned that the value of the slope of the sin at a point is just the value of the cosine at that point. The opposite is true for a cosine, except for a minus sine. In other words

$$\frac{dcos(\theta)}{d\theta} = -sin(\theta)\ and\ \frac{dsin(\theta)}{d\theta} = cos(\theta)$$

I also learned a couple of other interesting results. One is that the derivative of the product of two functions such as $y(\theta)x(\theta)$ is

$$\frac{dy(\theta)x(\theta)}{d\theta} = y(\theta)dx(\theta)/d\theta + x(\theta)dy(\theta)/d\theta$$

The other is that when θ is very small $sin(\theta) \sim \theta - \theta^3/6$ and $cos(\theta) \sim 1 - \theta^2/2$

I'm very interested in where these expressions come from and I intend to find out, but right now I don't have the time to take a deep dive so I'm just going to accept them.

If I call S(x, t) the derivative of A(x, t) then the slope S is

S = $(4\pi f\alpha/v)\cos(2\pi fx/v)cos(2\pi ft)$

I have mixed feelings about what I've accomplished. On the one hand I'm really excited to show this to Dad—it's been a couple of weeks of hard work, but the excitement of progress makes the effort more than worthwhile. On the other hand, although I calculated an expression for the derivative of A(x, t), I used the derivatives of sin and cos to do it, and I don't know how those relationships were obtained. So all I did was substitute one unknown for another. I actually took the calculations one level deeper by proving the relations for the derivative of sin and cos, but to do that I used the small angle approximations—and I don't know where those came from. That's something I'm eager to find out because I think, or at least hope, that it will finally get me to the deepest level of understanding. For the moment, however I've just accepted them. Here's how I determined that

$$\frac{dcos(\theta)}{d\theta} = -sin(\theta)$$

By definition $\frac{dcos(\theta)}{d\theta} = \lim\limits_{\Delta\theta \to 0}[cos(\theta + \Delta\theta) - cos(\theta)]/\Delta\theta$

It's interesting that the value of this function at $\theta = 0$ is 0/0, which seems to me to be meaningless, whereas I found that the limit itself is well defined. That relationship—between the value of a function at some point, and the value of the limit at the same point—is something else I intend to delve into more deeply, but in the meantime here's my proof that the limit exists.

Since cos (a + b) = cos(a) cos(b) - sin(a)sin(b) then

$$\lim\limits_{\Delta\theta \to 0}[cos(\theta + \Delta\theta) - cos(\theta)]/\Delta\theta = \lim\limits_{\Delta\theta \to 0}[cos(\theta)(cos(\Delta\theta - 1)) - sin(\theta)sin(\Delta\theta)]/\Delta\theta$$

Now using the small angle approximations, the result; viz the derivative of cosine is - the sin, follows in one step. I used a completely analogous procedure to find the derivative the sin.

Conversation 3

$$\tau = r_{cm}F_g \sin\theta$$

$$\omega = \frac{\tau}{L} = \frac{r_{cm}mg \sin\theta}{I\omega_s}$$

$$\theta = \sin^{-1}\frac{I\omega_p\omega_s}{mgr_{cm}}$$

Conversation 6

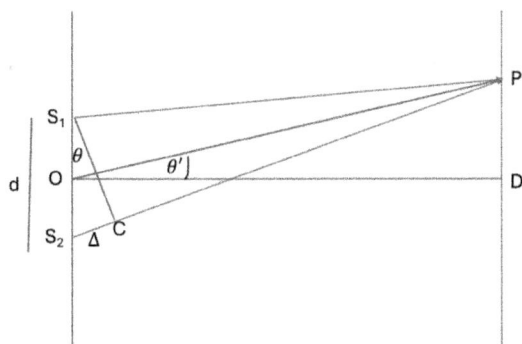

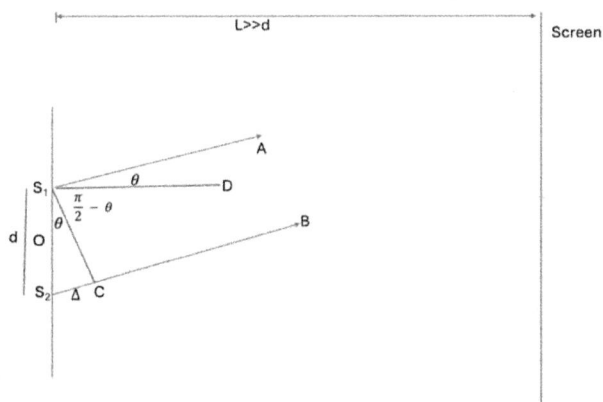

Uncle Zach showed me a sketch with multiple peaks of decreasing amplitude on both sides of a central maximum—he called them peaks of order one, order two, order three etc, as they diminish in magnitude. The question is what is the condition for constructive interference at these peaks? Δ in the figure illustrating Thomas Young's double slit experiment is the difference in path length between S_1P and S_2P and it must equal one wavelength at the peak of order one is to be constructive. Since $\Delta = d\,sin(\theta)$, the condition for constructive interference where these first peaks appear is $\lambda = d\,sin(\theta)$. This result confused me at first because it looks the same as the condition for destructive interference, but then I realized that the condition for destructive interference referred to two waves that started out a distance d / 2 apart.

The real problem is that θ isn't measurable, but the difficulty can be avoided if θ' which is measurable, can be used as a substitute for θ. Uncle Zach threw a curveball when he said the relationship depends upon θ and θ' being equal because I tried in vain to prove it. He then realized he had misled me and meant to say that θ' can be used, with negligible error, as a substitute for θ.

The essential equivalence became obvious once I recognized that S_1P and S_2P are essentially parallel in the regions of interest, which would be within tens or even hundreds of wavelengths from their points of origin. That would make $\sphericalangle AS_1C \sim \pi/2$ and it would then follow immediately that $\sphericalangle AS_1D = \sphericalangle S_2S_1C = \theta$.

Conversation 8

If I let A be the blade area in units of square meters, and let v be the wind velocity in units of meters per second, then their product, Av, which has units of m^3/s is the volume of air per unit time swept by the blades.

Now what? Since power is kinetic energy per unit time, I need to convert volume of air to mass of air, and I can do that by multiplying by ρ the density of air in Kg/m^3 so $Av\rho$ is the mass of air per unit time hitting the blades. Now it seems that the result I want can be obtained by multiplying $Av\rho$ by the kinetic energy per unit mass, $v^2/2$. That gives the power of wind swept by the blades as

$$P_{wind} = \frac{1}{2}Av^3\rho$$

This is one of the places where Uncle Zack helped me by explaining something very fundamental: all conversions between different forms of non-thermal energy have limited efficiency, and in particular the fraction of wind power, C_p, that can be captured by the turbine is less than 0.59, the so-called Betz limit. He said that 59% assumes idealized conditions: turbines operate below the limit because energy is lost to friction and turbulence. In addition, air must continue moving behind the turbine to allow new air to flow through. If all the energy were extracted, the air would stagnate, blocking the turbine from capturing energy.

And yes, he reminded me semi sarcastically, the limit is named after a scientist, the German physicist Albert Betz, who derived the equation enabling the estimate to be made.

So that makes it very easy to write down an expression for rotational power: taking the efficiency limitation into account the wind power transferred to the rotating blades is

$$P_{rot} = C_p \, P_{wind} = \frac{1}{2}C_p\rho Av^3$$

References

Conversation 1 Blood, Straws and Pressure

1. Bloomfield, L **How Everything Works**. 2001 John Wiley and Sons, *pp 130-139*

2. Goldberg, Stephen Clinical Physiology Made Ridiculously Simple: Color Edition 3rd Edition by MedMaster Inc, 2023

3. Halliday, D., Resnick, R., & Walker, J. (2013). Fundamentals of Physics. New York: Wiley. This introductory physics text includes a discussion of fluid statics and barometric pressure. It explains how atmospheric pressure and pressure differentials are crucial to understanding how fluids behave in closed systems, such as in straws.

4. Sear's and Zemansky's University Physics, with modern physics.-12th ed. Hugh D. Young, Roger A. Freedman; contributing author, A. Lewis Ford. This has been a standard introductory physics text for science and engineering students for more than six decades. It is fully up to date with extensive coverage of modern physics as well as the standard topics of mechanics electromagnetism, optics and thermodynamics.

5. West, John B. "Torricelli and the Ocean of Air: The First Measurement of Barometric Pressure", Physiology. Published 1 March 2013 Vol. 28 no. 2, 66-73 DOI: 10.1152/physiol.00053.2012. West describes Torricelli's groundbreaking experiments and how he arrived at the conclusion that "We live submerged at the bottom of an ocean of the element air, which by unquestioned experiments is known to have weight."

Conversation 2 Timbre and the Sound of Music

1. Békésy, Georg von, "The Ear", Scientific American 1957 vol 197, issue 2. The former physicist George Von Bekesy explains his pioneering research on the mechanics of the ear, for which he was awarded the 1961 Nobel Prize in Physiology or Medicine.

2. Deutsch, J., Clarkson, "Nature of the Vibrato and the Control Loop in Singing". Nature 183, 167–168 (1959). https://doi.org/10.1038/183167a0. Deutsch investigates the control loop in singing, emphasizing the auditory feedback mechanism. Experiments with delayed auditory feedback demonstrate that singers rely heavily on hearing their voice for frequency control. A delay increases both the time and amplitude of pitch fluctuations. Findings suggest that corrections to errors in pitch occur with very short reaction times (around 0.1 seconds) and at pitch deviations below the typical discrimination threshold.

3. De Wesendonk, C. "Nature of Vowel Sounds", Nature 107, 12 (1921).

   https://doi.org/10.1038/107012c0. De Wesendonk discusses the nature of vowel sounds, their acoustic properties, production mechanisms, and significance in speech. Key points include production of vowel sounds when air flows through the vocal tract without significant constriction; how the shape and positioning of the tongue, lips, and jaw determine the vowel quality, the distinct formant frequencies, which are resonant frequencies of the vocal tract which help distinguish between different vowel sounds (e.g., 'a', 'e', 'i', 'o', 'u').

4. Hart, Gerald W. Harmonies and Discords: The Science of Music. Oxford University Press, 2011.

5. Levitin, Daniel J This is your brain on music Penguin Press , 2006. Levitin explores the cognitive neuroscience of music, examining how the brain processes, understands, and enjoys music. He introduces readers to fundamental concepts about music and its elements, such as pitch, tempo, rhythm, harmony, and timbre and explains how these components interact to create the rich and diverse experiences we associate with music. He also provides a foundation for understanding how the brain processes music, highlighting its complexity and the interconnectedness of neural pathways involved in auditory perception, memory, and emotion.

6. Roederer, J. G. (2008). The Physics and Psychophysics of Music: An Introduction. This book provides an engaging explanation of how the human auditory system detects and processes sound. It discusses the remarkable sensitivity of the ear to pressure changes and its ability to differentiate frequencies, which underpins our perception of music and speech.

7. Rossing, T D, F. R. Moore and P.A. Wheeler (2002), The Science of Sound. 3rd edition Addison-Wesley. The authors provide a comprehensive overview of sound physics, including the formation of standing waves and the significance of resonance in musical instruments. The text highlights how harmonic frequencies interact to create the unique sound of each instrument An excellent

blend of the physics of sound at an elementary level with various aspects of music including pitch and timbre, tone and harmony and acoustic characteristics of string and brass instruments

8. Wilson, Herbert A. (1962) "Sonic Boom", Scientific American, Vol. 206, No. 1, pp 36-43. Wilson explains how sonic booms arise from the compression of air pressure waves that coalesce into a shock wave, forming a conical pattern called the Mach cone. The article also addresses the environmental and societal challenges posed by sonic booms, such as structural damage to buildings, noise pollution, and disruption to communities. He underscores the need to balance the benefits of supersonic travel with its adverse consequences, emphasizing the importance of further scientific understanding and technological innovation to minimize the boom's impact while enabling advances in high-speed aviation.

9. Yost, W. A. (2015). Fundamentals of Hearing: An Introduction. 5th edition Academic Press. This text explains the mechanics of hearing, including the detection of minute air pressure variations by the tympanic membrane and the auditory system's function in converting these variations into interpretable electrical signals for the brain.

Conversation 3 Planes, Tops, and Gravity Defiance

1. Anderson, J. D. (1999). Introduction to Flight. New York: McGraw Hill. This introductory text, which was written for engineering students, explains the principles of aerodynamics, including the generation of lift and the role of airfoil shapes in creating pressure differences

2. Babinsky, Holger "How Do Wings Work?" ,www.iop.org/journals/physed. Holger Babinsky examines the common misconceptions surrounding the generation of lift by wings. He critiques the widely held belief that lift is primarily due to the air traveling faster over the curved upper surface of the wing, resulting in lower pressure, as explained by Bernoulli's principle. Instead, he emphasizes the importance of the angle of attack—the angle between the wing and the oncoming airflow—in creating lift. He explains that a wing at a positive angle of attack deflects air downwards, and according to Newton's third law, this downward deflection results in an upward reaction force, which we perceive as lift.

3. Benyus, J. M. (1997). Biomimicry: Innovation Inspired by Nature. New York: Harper Perennial. Benyus explores how biological systems inspire engineering solutions, including flight technology

modeled on birds. The book highlights how principles from nature lead to innovations in sustainability and design.

4. Klein, F and A Sommerfeld, The Theory of the Top, *Volume IV, Astronomical and Geophysical Applications*, and *Technical Applications of the Theory of the Top.* Springer Nature.

5. Kleppner, D., & Kolenkow, R. (2013). An Introduction to Mechanics. Cambridge: Cambridge University Press. While primarily a 2nd year college level textbook, this resource provides an intuitive explanation of angular momentum, precession, and the dynamics of spinning objects. It offers insights into how gyroscopic motion is used in practical applications such as navigation and stabilization.

6. Savitsky, Zack, **The American Institute of Physics (AIP)**, (2015). *The Wondrous Physics of Spinning Tops*, Inside Science. https://ww2.aip.org/inside-science/the-wondrous-physics-of-spinning-tops A qualitative reflection on the universal laws that governs rotating objects and precession, from spinning tops to neutron stars

7. Precession of Spinning Top http://hyperphysics.phy-astr.gsu.edu/hbase/top.html

8. Feynman, R.P, R. Leighton and M. Sands The Feynman Lectures, v 1, chapter 20, Addison Wesley, Menlo Park CA.

Conversation 4   Color, Rainbows and Mirages

1. Zimmer, Carl. *She Has Her Mother's Laugh: The Powers, Perversions, and Potential of Heredity.* Dutton, An Imprint of Penguin Random House, 2018. This article explores how heredity influences physical traits, including skin color, and how these traits evolved as human populations adapted to different environments. It explains the role of melanin in protecting against UV radiation and discusses how genetic adaptations to latitude and climate have influenced skin color distribution worldwide.

2. Gibbons, Ann "Why do Europeans have white skin", Science, April 2, 2015. The article highlights genetic studies indicating that early European hunter-gatherers had darker skin, and the lighter skin pigmentation prevalent today became widespread only within the last 5,000 to 7,000 years. This change is associated with the spread of agriculture and dietary shifts, which may have influenced the need for increased vitamin D synthesis in the skin. The more general topic of skin color is covered extensively in

   https://en.wikipedia.org/wiki/Human_skin_color?utm_source=chatgpt.com

3. Phil Plait, "Why does smoke turn the sky orange", Scientific American June 16, 2023,

4. Minnaert, M.G.J. (1993) "Light and Color in the Outdoors" Springer Verlag. This is a classic resource that describes and explains the various optical phenomena including the shape of the dapples beneath a tree on a sunny day, rainbows, mirages, and haloes in clear language accessible to the layman.

5. Nassau, Kurt, "The Causes of Color", Scientific American 243(4), 124-151,1980

6. Fraser Alistair B. and William H. Mach "Mirages" Scientific American Jan 1976 p 104-111. The authors describe various types of mirages and the physics of their formation.

Conversation 5   Light: We See So Little

1. Hubilla Conelisa N "Why Can't We See Ultraviolet Light? Exploring the Limits of Human Vision" The Science Times Apr 29, 2024.

   https://www.sciencetimes.com/articles/49931/20240429/why-t-see-ultraviolet-light-exploring-limits-human-vision.htm

   This is a short elementary explanation of the ability of humans to see ultraviolet light, which our lenses filter to protect our eyes. While young people can perceive some UV light, this ability declines with age. Many animals, however, can see UV light, which helps them with tasks like hunting, navigation, and communication.

2. Yong, E. An Immense World Random House, 2022. Yong emphasizes the concept of the "umwelt" to describe an organism's unique sensory world, and the vast and varied ways animals perceive their environments through senses such as echolocation, ultraviolet vision, and electric fields. He argues that the human-centric view of the world limits our understanding of nature and urges us to appreciate the diversity of sensory experiences that shape life on Earth.

3. Hawking, S. (2001). The Universe in a Nutshell, Bantam Books. This book includes a lucid description of the speed of light as a universal constant. Hawking explains how light can be understood as both a wave and a particle, and how this duality is essential to our understanding of the Universe. While the book doesn't delve deeply into the specific details of electromagnetic waves, it provides a comprehensive overview of the fundamental principles of physics, including the nature of light and its connection to the electromagnetic spectrum.

4. Jennifer Lombardo "Lightning" (Scientific American Investigates Things That Light Up the Sky) December 30, 2024, https://mitpressbookstore.mit.edu/book/9781725352162 In one of a series of books by Scientific American Educational Publishing, Jennifer Lombardo discusses the immense power of lightning which can turn sand into glass, boil water, and cause trees to explode. Also discusses cultural myths and legends about electrical storms. Excellent graphics.

Conversation 6 Seeing Through Objects

1. Altman, Lawrence, New York Times, 12 October 1979 Section A1. American and Briton Get Nobel Prize for X-Ray Advance.

2. Born M and F Wolf, Principles of Optics. The classic book on optics. Suitable for advanced undergraduate physics and engineering students.

3. "Wave-Particle Duality of Light Phenomena", September 2024, Carolina Knowledge Center. This article provides an overview of the wave-particle duality of light, a fundamental concept in quantum mechanics. It discusses historical experiments, such as the photoelectric effect, that demonstrate light's dual nature. The piece serves as an educational resource, explaining how light

can exhibit both wave-like and particle-like properties and the implications of this duality in modern physics.

4. **"Transparent, Opaque, and Translucent Objects",** CK-12 Foundation. This article explores how light interacts with matter through reflection, transmission, and absorption. It explains the differences between transparent, translucent, and opaque materials, providing clear examples and illustrations to enhance understanding. The content is designed to be accessible to readers without a scientific background.

5. Encyclopedia Britannica

   https://www.britannica.com/science/quantum-mechanics-physics/Einstein-and-the-photoelectric-effect.

   This article provides an elementary introduction to the physics of the photoelectric effect and how it demonstrated the particle nature of light while preserving its wave-like behavior in certain experiments. This duality laid the foundation for quantum mechanics and helped explain phenomena such as diffraction and interference.

6. Feynman, R. P., Leighton, R. B., & Sands, M. (1963). The Feynman Lectures on Physics, Vol. 1 Addison-Wesley. This foundational text offers an in-depth introduction to fundamental principles of physics, including mechanics, radiation, and heat. Authored by renowned physicist Richard Feynman and his colleagues, the lectures provide clear explanations and insights into the behavior of light, the photoelectric effect, and the development of quantum mechanics.

7. Franklin, R. E. (1953). "Structure of DNA from X-ray Evidence." Nature. Rosalind Franklin's work demonstrates how X-ray crystallography provides detailed insights into molecular structures. The article discusses the techniques used to capture diffraction patterns and their interpretation.

8. Kittel, C American Journal of Physics, 1968 In his 1968 article "X-Ray Diffraction from Helices: Structure of DNA," a prominent solid state physicist examines the application of x-ray diffraction techniques to helical structures, with a focus on DNA. He discusses how the helical arrangement of DNA influences its diffraction patterns and explores the methods used to interpret these patterns to deduce the molecular structure. Kittel's analysis provides insights into the principles of x-ray diffraction as applied to complex biological molecules, highlighting the interplay between physics and biology in understanding the structural properties of DNA.

9. Orzel, Chad "The optics of superman's X-ray vision", Forbes Mar 25, 2016. Physicist Chad Orzel examines the scientific plausibility of Superman's x-ray vision. Orzel explains that for Superman to see through objects using x-rays, his eyes would need to emit and detect x-ray radiation. However, x-rays interact with matter differently than visible light; they tend to pass through substances rather than reflecting off surfaces. This characteristic makes it challenging to form clear images based on reflected x-rays. Additionally, the high energy of x-rays poses safety concerns, as excessive exposure can be harmful to living tissues. Orzel concludes that, while intriguing, the concept of x-ray vision as depicted in comics is not feasible with our current understanding of physics.

10. Watson, J. D., & Crick, F. H. C. (1953). "DNA and X-ray Diffraction." Nature. Watson and Crick's seminal paper explains how X-ray diffraction patterns were used to deduce the double-helix structure of DNA. The article underscores the critical role of diffraction in molecular biology.

Conversation 7 The Earth's Changing Climate

1. Alley, R. B. (2014). The Two-Mile Time Machine Ice Cores, Abrupt Climate Change and our Future, Princeton University Press. Alley discusses the role of natural climate variability within long-term warming trends. The book uses ice core data to demonstrate the significance of decadal and centennial patterns in the context of global climate.

2. Archer, D. (2009). The Long Thaw: How Humans Are Changing the Next 100,000 Years of Earth's Climate. Princeton: Princeton University Press. Summary: Archer compares natural processes, such as volcanic eruptions and orbital cycles, with the rapid warming caused by human emissions and explores the profound and long-term consequences of human activity on Earth's climate. He argues that the effects of fossil fuel burning and carbon dioxide emissions are not temporary but will resonate across millennia, far outlasting the human civilizations that caused them. By weaving together scientific insights and accessible language, Archer makes a compelling case for urgent action to mitigate climate change to preserve the planet for future generations

3. Dressler, Andrew Introduction to Modern Climate Change, Cambridge University Press New York 2016. This book provides an introductory resource for understanding the complexities of modern climate change. Written for both students and general audiences, Dressler addresses the complexity of the climate system, including greenhouse gases, feedback mechanisms, and energy

balance, while also discussing mitigation strategies, such as renewable energy adoption and policy frameworks. With clear explanations and supporting data, the book emphasizes the importance of informed decision-making to address the challenges posed by climate change.

4. Hansen, J. (2009). Storms of My Grandchildren. New York: Bloomsbury. James Hansen, a leading climate scientist, offers a compelling and urgent warning about the future of Earth's climate by explaining the science of global warming and the catastrophic consequences of inaction. Hansen emphasizes the role of fossil fuels in driving climate change and critiques political and corporate interests that have hindered meaningful action. Through personal anecdotes and scientific insights, he presents a call to action for policymakers and the public to make bold changes to ensure a sustainable future for future generations. The book serves as both a scientific explanation and a passionate plea for immediate and transformative climate action.

5. Hansen, J., Sato, M., & Ruedy, R. (2012). "Perception of Climate Change." Proceedings of the National Academy of Sciences of the United States of America, 109(37), E2415–E2423. In this influential article, Hansen and colleagues present a comprehensive analysis of the increasing frequency and intensity of extreme weather events linked to climate change. By analyzing global temperature records and regional weather patterns, the authors demonstrate a clear shift in the climate system, showing how human-induced global warming has altered the likelihood of extreme weather occurrences. They argue that the perception of climate change is no longer confined to abstract temperature trends but is evident in the tangible experience of extreme heat waves, droughts, and other significant events.

6. Keeling, C. D. (1960). *Tellus*, 12(2), 200–203. "The Concentration and Isotopic Abundances of Carbon Dioxide in the Atmosphere." This foundational paper introduces the Keeling Curve, documenting the rise of CO2 concentrations over time. It explains how industrialization has disrupted natural carbon cycles, leading to unprecedented CO2 levels. Keeling presents the first comprehensive measurements of atmospheric carbon dioxide (CO_2) concentrations and their isotopic compositions.

7. Kolbert, E. (2014). The Sixth Extinction: An Unnatural History. Henry Holt and Company. This Pulitzer Prize-winning book argues that Earth is undergoing a sixth mass extinction, driven largely by human activities.

8. Lee Kump, "The Last Great Global Warming," Scientific American July 2011.

9. What's the Difference Between Climate and Weather.

10. https://www.noaa.gov/explainers/what-s-difference-between-climate-and-weather?utm_source=chatgpt.com. A brief introduction from the National Oceanographic and Atmospheric Administration.

Conversation 8 Clean energy and the Creation of Electricity

1. Lewis, N. S., & Nocera, D. G. (2006). "Powering the Planet: Chemical Challenges in Solar Energy Utilization." Proceedings of the National Academy of Sciences. This paper explores advancements in solar energy technology and the challenges of efficiently capturing and storing solar power. The authors advocate for integrating solar energy into a sustainable global energy strategy.

2. MacKay, D. J. C. (2009). Sustainable Energy: Without the Hot Air. Cambridge: UIT Cambridge. MacKay offers a clear and quantitative examination of energy sources, including nuclear power. He explains its potential to generate large amounts of low-carbon energy and addresses common concerns about safety and waste management.

3. McCully, P. 12 November 2010 Silenced Rivers: The Ecology and Politics of Large Dams.

4. Sioshansi, F. P. (2011). Smart Grid: Integrating Renewable, Distributed & Efficient Energy. Boston: Academic Press. Sioshansi discusses the modernization of electricity grids to accommodate renewable energy and improve efficiency. He highlights the critical role of advanced energy storage systems in balancing supply and demand in the transition to clean energy.

5. Smil, V. (2017). Energy and Civilization: A History. Cambridge: MIT Press. Smil examines the technologies used to transmit electricity over long distances, including step-up and step-down transformers. He also discusses energy storage challenges and the importance of developing efficient batteries.

6. **"PV Cells 101: A Primer on the Solar Photovoltaic Cell"** U.S. Department of Energy *Source: Energy.gov,* August 10, 2018 This article provides an accessible introduction to how photovoltaic

(PV) cells convert sunlight into electricity. It explains the basic principles of the photovoltaic effect, the materials used in PV cells, and the advancements in solar technology over the past two decades. The piece is designed to help readers understand the fundamentals of solar energy conversion. U.S. Department of Energy

Conversation 9: Artificial Intelligence (AI)

1. Bear, M. F., Connors, B. W., & Paradiso, M. A. (2020). Neuroscience: Exploring the Brain. Philadelphia: Wolters Kluwer. a widely used textbook that provides a comprehensive overview of neuroscience. Known for its clear explanations, engaging writing style, and numerous illustrations.

2. Bengio, Yoshua, Scientific American, June. 2016 pp 46-51. "Machines Who Learn". Bengio explains how machine learning allows computers to improve performance on tasks by analyzing data rather than being explicitly programmed. He highlights the resurgence of deep neural networks, which mimic the brain's interconnected neurons; applications including speech recognition, image analysis, and natural language processing, where machine learning has led to breakthroughs; learning by recognizing patterns and adapting to data without human intervention, and explores how machine learning could revolutionize industries for better and worse.

3. Brynjolfsson, E., & McAfee, A. (2014). The Second Machine Age: Work, Progress, and Prosperity in a Time of Brilliant Technologies. New York: W.W. Norton. The authors examine how AI and automation are reshaping the economy, creating opportunities for innovation while disrupting traditional labor markets. They emphasize the importance of preparing for these changes through education and policy.

4. **Russell, Stuart** (2016). "*Should we Fear Supersmart Robots?*"**June**, Vol. **314**, pp. **58-59. Russell discusses principles to ensure guardrails against misalignment of our commands and an action taken by an AI.**

5. LeCun, Y., Bengio, Y., & Hinton, G. (2015). "Deep Learning." Nature. This article provides an overview of artificial neural networks and their inspiration from biological systems. The authors discuss how artificial networks process data and adapt through training, drawing parallels to the functioning of biological neurons.

6. Kaufman, Claire, "Unravelling the Complexity of the Human Brain: A Comprehensive Study of Anatomical Structures and Functions". J Morphol Anat 7 (2023): 286 Claire Kaufman provides a comprehensive overview of the human brain's anatomy and functions. The article delves into the intricate network of neurons and their connections, highlighting the crucial role of understanding this architecture in advancing neuroscience research and developing treatments for neurological disorders. It discusses the various regions of the brain, their specific functions, and how they interact to produce complex behaviors and cognitive processes. The author emphasizes the importance of ongoing research and technological advancements in unraveling the mysteries of the human brain. As our understanding of the brain deepens, we can expect to make significant strides in treating neurological diseases and disorders, as well as enhancing our understanding of human consciousness and cognition.

7. Tegmark, M. (2017). Life 3.0: Being Human in the Age of Artificial Intelligence. New York: Knopf. Tegmark explores the transformative potential of AI, addressing both its benefits, such as advancements in healthcare and automation, and challenges, like job displacement and ethical concerns. He advocates for proactive governance to ensure AI's responsible development.

8. Seung, S. (2012). Connectome: How the Brain's Wiring Makes Us Who We Are. New York: Houghton Mifflin Harcourt. Seung explores the brain's parallel processing capabilities through its dense neural networks. He contrasts this with the sequential nature of most computers, emphasizing the brain's adaptability and complexity. He argues that our thoughts, memories, and personality are not solely determined by our genes but are shaped by the unique patterns of connections between neurons, known as the connectome. Seung discusses how our experiences and learning shape our brain's wiring, and how this wiring, in turn, influences our behavior and cognition. He also explores the ethical implications of understanding and potentially manipulating the human connectome.

9. *Vasqani, A., Shazeer, N; ; Parmar, Niki; Uszkoreit, Jakob; Jones, Llion et al (2017). Attention is All you Need. Advances in Neural Information Processing Systems. 30. Curran Associates, Inc. An important advance in transformer architecture that underlies current bots.*

Index

A

action potential	212, 214
acceleration	9, 16, 105
sonic boom	105
alternating current	113
aerosols	107, 176-178
atmospheric cleansing	180
atomic number	112, 113

B

barometer	7,15
Bernoulli, Daniel	59
Bernoulli principle	59,63
family	64
Biot-Savart law	110,112
Blackbody radiation	162,163

C

carbon cycle	159
chemical bonds	175
clean energy	180-184
climate sensitivity	177
cone cells	83
Crick, Frances	135,145

D

da Vinci, Leonardo	66
decibel, definition	37-40
rock concert	41-42
density	
air properties	8
optical	92-93
mercury	8,9,20

water	7,12,14-16,20,35,78,89,91,94-96,113-114,114-116, 143,161,172,175,180,198,237
stellar	38-39
and lift	61-63,65,190
and refraction	27, 91-93, 95-96
diffraction	135, 137-139, 143-145
DNA	78, 118,132-135, 137, 144-145, 158, 233

E

Einstein, Albert	43,100,118,125-126,129,131,135,143
electricity and magnetism	110,112,113,116
electromagnetic force (EMF)	187-188,195
electromagnetic waves	29, 43, 97, 110, 114-115, 124, 147
electromagnetic quanta	125,129
energy	
energy imbalance	167-168
planetary	177
evolution	
auditory sensitivity	39
and morphology	25
stellar	43
and engineering	65-66

F

Faraday, Michael	112-113, 187
Faraday's law	188
feedback, Earth system	149, 150, 175-177
windmill blades	190
Volterra - Lotka	149-150
flight	
lift	61-66, 182, 190
drag	61-62, 190
planes	59-63,65,73,190
birds	59-60, 65-66
angle of attack	61-64, 66
fossil fuels	158,163,173-174,184-186
frequency	
sound waves	35,38,42,100,105-106,114-115
light	27,43
and red giant	45

G

geological processes   168

geophysics, geochemisty   165, 184

glacial cycles   84,169,170-171

gravity   1,15,43
        of the moon   5,208
        defiance   15
        general relativity   36,39
        star formation   45,46
        Newton's Law   207

gravitational waves   22,42-43,46

Gray   78

greenhouse gases   71,157-164,167-168,170,173,176-178,180,184,187

H

hearing   37
        frequency dependence   37-38
        threshold   41

Hertz, Heinrich   29,125

hertz(HZ)   29,55

Huygens, Hans Christiaan   138
        Huygens' principle   138

hydroelectric power   114

I

ice core   172

ice sheets   84,170,172,178

industrial revolution   157-158,163,168,174-175

infrared radiation   99-100, 102,109,160,175\

interference   136, 140
        light waves   139-140,142-143,146

ionosphere   107
        radio wave reflection   107

K

Keeling, Charles   164-165

kinetic energy   32-33
        of turbines   113, 242
        rotational   113,187,189
        gravitational   111

Kuhn, Thomas | 125, 130

L

Land management | 180
    And shock waves | 105
light, visible | 43, 82, 85, 97, 99, 107, 109, 114, 118, 121-122. 124
                                       129, 133, 137, 146
    speed | see electromagnetic
Los Alamos | 42, 56, 62, 80, 87, 139-140

M

magnetism | see electricity and magnetism
magnetic field | 110-114, 137, 188-189, 193, 195
mass and weight distinguished | 2
matter wave | 27, 42, 46
Maxwell, James Clerk | 112, 116
    Maxwell's Theory | 113
mechanical energy | 137, 157-158, 187, 198
    windmill blades | 182
melanin | 84-86
microwave radiation | 116
    oven | 116
middle C | 29, 30, 38-39, 47, 49, 54-56, 169
Milankovich, Milutin | see orbital parameters
mirages | 80, 91, 93-95
Muller, Herman | 76
    X-rays and mutations | 76

N

neuron | 202, 212-214, 219, 223-224, 226
neutron star | 46
Newton, Isaac | see gravity
    third law | 59, 64, 66, 182

O

optical density | 92
optical illusion | 95
orbital parameters | 170
Ørsted, Hans Christian | 95, 96

P

paradigm shift   see Thomas Kuhn

Pascal, Blaise   8

photoelectric effect   118, 125-126, 129-131, 199

photoreceptor   215

photovoltaic
     effect   199
     cells   199, 202

photon   124-125, 131-132, 143-144, 146, 199, 202, 217

piano
     strings   23, 30, 33, 57, 169
     length frequency relation   30, 47-49, 55

Planck, Max   118
     quantum hypothesis   125
     and Einstein   125
     blackbody spectrum   162-163

plasma frequency   107

plate tectonics   156

Planetary temperature   72, 149, 175, 178, 184
     Cenozoic era   155

predator prey dynamics   149-150

pressure
     atmospheric   6-10, 15-16, 19-20, 88,237
     blood   1, 2, 7, 19-21
     hydrotastic in a tube   15-17
     arterial   20, 135, 137, 145, 158-158, 219, 221-223, 226

protein   118, 125, 191

quanta   125, 129

quantum mechanics   27, 46, 129, 132

R

radio waves   77, 97, 99-102, 107-109, 114

rainbows   89, 91, 95

rational number

Rayleigh, John W S
     scattering   82
     optical resolution criterion   142

refraction   27
     rainbow   89, 96
     mirage   91, 93, 95

reflection   27, 29, 42, 96
     and plasma frequency   107

and atomic structure	133
resonance	57-58, 132, 146
retinal ganglion	215
Roentgen, Wilhelm	76
unit definition	77
X-rays	125, 131, 133, 145

S

Shakespeare, William	5, 144
shock wave	105
Sievert	78
sky color	
evening	82-83
solar constant	162
sound	23
propagation in air	32
in water	35
sound speed vs altitude in air	60
special relativity	100, 111
spinning top	
angular momentum	67, 71
precession	68-69
Stefan-Boltzmann constant	163
Superman	118-121

T

Temperature cycles	
glacial, inter-glacial	164, 155
physiological	98
thermal equilibrium	161
thunder	103-105
torque	67, 69, 71-74
Toricelli, Evangelista	7-10
transparency, and atomic structure	146
tuning fork	28-30, 32, 33, 47, 53
Turbines	
wind	
efficiency, blade power	189, 191
water	198
tip speed ratio	192

U

ultraviolet light	85, 114, 125
units	4, 5, 7, 10, 16, 20, 24, 30, 37, 68, 82, 125, 191, 200 226, 236
SI	8-9, 78, 101, 236
imperial	236

V

vibrato	22, 58
vocal system anatomy	56-57

W

Watson James	135, 145
wavelength	34-35, 47-48, 53, 82, 84, 100-102, 107-109, 123, 136, 138 142, 146, 238, 241
water waves	35
wave-particle duality	118
many worlds interpretation	143
waves	
standing	53
nodes	48, 50, 53-54
weather vs climate	147, 160
weathering, chemical	149-150
World Health Organization	152, 175

X

X-rays	76-78, 97, 99, 121, 123-124, 131-133, 135, 137. 145, 146

Y

Young, Thomas	135
double slit experiment	136, 138, 228

www.ingramcontent.com/pod-product-compliance
Lightning Source LLC
Chambersburg PA
CBHW081807200326
41597CB00023B/4175